CLIMATE CHANGE AND ITS CAUSES, EFFECTS AND PREDICTION

GLOBAL CLIMATE CHANGE AND UNITED STATES SECURITY INTERESTS

CLIMATE CHANGE AND ITS CAUSES, EFFECTS AND PREDICTION

Additional books in this series can be found on Nova's website under the Series tab.

Additional E-books in this series can be found on Nova's website under the E-book tab.

DEFENSE, SECURITY AND STRATEGIES

Additional books in this series can be found on Nova's website under the Series tab.

Additional E-books in this series can be found on Nova's website under the E-book tab.

CLIMATE CHANGE AND ITS CAUSES, EFFECTS AND PREDICTION

GLOBAL CLIMATE CHANGE AND UNITED STATES SECURITY INTERESTS

ANGELO C. DISTEFANO
AND
MARY HURLEY
EDITORS

Nova Science Publishers, Inc.
New York

For permission to use material from this book please contact us:
Telephone 631-231-7269; Fax 631-231-8175
Web Site: http://www.novapublishers.com

Library of Congress Cataloging-in-Publication Data

ISBN: 978-1-62081-288-4

Published by Nova Science Publishers, Inc. † New York

Contents

PREFACE

Changes in climate patterns and their impact on the physical environment can create profound effects on populations in parts of the world and present new challenges to global security and stability. Failure to anticipate and mitigate these changes increases the threat of more failed states with the instabilities and potential for conflict inherent in such failures. Because of the increasing importance of climate change on U.S security, this book examines the need to adapt, manage, and mitigate these consequences. An overview of the impacts of climate change over recent decades is discussed with a focus on the political consequences of climate change as it relates to national and international security, with special attention to the African continent due to the vulnerability of African nations with high potential to intersect with U.S. national interests.

Chapter 1 - This report describes observable climate change and its consequences. It does not attempt to address the complex and controversial set of causes, nor does it offer recommendations on the possibility of changing the pace or scope of climate change. Instead, the focus is on the need to manage consequences. To that end, this report describes evidence of impacts over recent decades while recognizing uncertainty about the pace of future changes. It discusses the shortfalls in climate information, climate science, and climate models and the resulting scope of the uncertainties in projections. At the same time, it provides compelling evidence that climate impacts are observable, measurable, real, and having both near and long-term consequences. It recognizes that changes already underway are having, and will continue to have, major consequences for the political, economic, and geographic world as we know it. This report provides a set of recommendations on approaches to adaptation and dealing with the near-term effects on populations that impact US and international security interests.

Chapter 2 - Climate change is a complex, crosscutting issue that poses risks to many existing environmental and economic systems, including agriculture, infrastructure, ecosystems, and human health. A 2009 assessment by the United States Global Change Research Program (USGCRP) found that climate-related changes—such as rising temperature and sea level—will combine with pollution, population growth, urbanization, and other social, economic, and environmental stresses to create larger impacts than from any of these factors alone.

According to the National Academies, USGCRP, and others, greenhouse gases already in the atmosphere will continue altering the climate system into the future, regardless of emissions control efforts. Therefore, adaptation—defined as adjustments to natural or human systems in response to actual or expected climate change—is an important part of the response to climate change.

This testimony addresses (1) the data challenges that federal, state, and local officials face in their efforts to adapt to a changing climate, (2) the actions federal agencies could take to help address these challenges, and (3) federal climate change strategic planning efforts. The information in this testimony is based on prior work, largely on GAO's recent reports on climate change adaptation (GAO-10-113) and federal climate change funding (GAO-11-317). These reports are based on, among other things, analysis of studies, site visits to areas pursuing adaptation efforts, and responses to a web-based questionnaire sent to federal, state, and local officials.

In: Global Climate Change ... ISBN: 978-1-62081-288-4
Editors: A. C. DiStefano and M. Hurley

Chapter 1

TRENDS AND IMPLICATIONS OF CLIMATE CHANGE FOR NATIONAL AND INTERNATIONAL SECURITY*

Defense Science Board

EXECUTIVE SUMMARY

Purpose

This report describes observable climate change and its consequences. It does not attempt to address the complex and controversial set of causes, nor does it offer recommendations on the possibility of changing the pace or scope of climate change. Instead, the focus is on the need to manage consequences. To that end, this report describes evidence of impacts over recent decades while recognizing uncertainty about the pace of future changes. It discusses the shortfalls in climate information, climate science, and climate models and the resulting scope of the uncertainties in projections. At the same time, it provides compelling evidence that climate impacts are observable, measurable, real, and having both near and long-term consequences. It recognizes that changes already underway are having, and will continue to have, major consequences for the political, economic, and geographic world as we know it.

* This is an edited, reformatted and augmented version of the Report of the Defense Science Board, dated October 2011.

This report provides a set of recommendations on approaches to adaptation and dealing with the near-term effects on populations that impact US and international security interests.

While the effects of climate change are uneven and subject to a complex set of influences, some long-term trends seem clear. For example, to the extent that greenhouse gases are causing land and sea surface warming, the long life of some greenhouse gases in the atmosphere mean that effects will continue to increase and will be long lasting, even with no further addition to the concentration of greenhouse gases (e.g., water vapor (H20), carbon dioxide ($C0_2$), nitrous oxide (N_20), methane (CH_4), and ozone (03)). Hence, this is a challenge that cannot be "solved". Instead, it must be managed for the long-term. This report addresses the need and prospects for approaches to address the near-term impact on human populations and longer-term adaptation to climate change. The impact on human populations, near- and long-term, translates to impact on national and international security.

Cycles of climate change are not new to the planet. But, an important difference today is that effects are exacerbated by the increased density of populations, particularly in those areas most vulnerable to climate change.

The effects can be further exacerbated by the nature of the human response or the lack of response.

While adaptation approaches must cover a wide spectrum of effects and consequences, the challenges associated with climate change generally fall into three interrelated classes of needs each with several subcomponents:

- Population support system resiliency
    - Water and food security
    - Health
    - Energy
- Human security
    - Population dislocation
    - Armed conflict
- Political continuity
    - Continuity of governance
    - Economic viability

Although recent data demonstrate an accelerating rate of climate change, the uncertainties about continued changes and populations' capabilities to adjust to those changes have led this report to a hedge strategy that recognizes the wide range of potential scenarios about the scope and rate of change. It focuses on climate related actions that will be beneficial to national and international security, regardless of the rate of climate change. This report places particularly strong emphasis on the need for programs and activities that provide better and more credible information to decision makers.

Climate Observation, Modeling, and Projection Needs

This report uses data and projections from a wide variety of sources to discuss trends and consequences. While this data comes from credible sources, climate information systems and climate modeling fidelity leave room for wide variances in projections. Hence, while the historical and recent trends seem clear, and consequences are visible in many parts of the world today, projections for future rates of change and impacts are far less clear.

Decision-making support demands a responsive, comprehensive, and informative climate information system. The system will need to include extensive, credible reporting and better science to predict the long-term implications of impacts and demand for adaptability. Progress towards producing such a system will require a systematic effort to define requirements, architecture, and implementation plans.

Currently no coherent, integrated climate information system capable of generating reliable, sustained, and actionable climate data and projections exists. Today's climate observations and models exist only as a loose federation of programs at many government agencies, academia, industry, and nongovernmental organizations (NG0s). While some of these assets are operational systems, the majority of observational assets and many of the modeling assets, are intended primarily for exploratory science rather than for supporting operational, long-term climate assessments. Many observations are intended to help improve basic climate process understanding, such as closing global-scale energy, carbon, and water budgets. They do not offer the space and time resolution, completeness, or accuracy to support needed improvements in, or validation of, regional climate models, particularly in developing countries.

The plethora of climate models present their own risk management challenges. Some of the models purport to provide highly accurate, long-term predictions (forty to one hundred years). For these predictions, there is a need for validation and verification standards. Also needed are clear uncertainty bounds and attention to the unpredictable variables that impact climate in the near-term. Few of the models purport to provide accurate predictions that cover the planning time frame typical of most government activity.

The observational systems that provide the needed climate data are deployed in the four traditional physical domains—land, sea, air, and space. Space-based systems that are funded, developed, deployed, and operated by multiple agencies will play a major role because they provide the real-time, continuous observational data required to monitor changes in the climate system. Given the need for comprehensive global data to understand current conditions and make better predictions about future changes, there is a need for a comprehensive approach to space-based systems and systems operating in other domains.

While this report approaches climate change from a global perspective, it places special emphasis on Africa. The combination of existing climate change and the vulnerability of African nations to such change warrants special attention. African vulnerabilities are the result of widespread subsistence farming, fragile governance, economic shortfalls, ethnic tensions, and other factors.

Neither the needs of the Department of Defense (DOD) for climate information systems and assessment processes, nor gaps relative to current capabilities and planned future capabilities have been rigorously studied to date. The US government needs a scientifically robust, sustained, and actionable climate information system that addresses these and other issues. The details on the needs of such a system, and key barriers to its establishment are described in Appendix A of this report.

Trends

This report addresses observed trends and some predictions based on these trends. It does not attempt to address the causes of climate change. The observable trends over multiple decades include:

- Increasing land and sea surface temperatures
- Changing ocean temperature
- Changing ocean chemistry (acidity and salinity-impact on ecosystems and circulation)
- Declining mass of Greenland and Antarctic ice sheets
- Declining glaciers and snow cover
- Decreasing and thinning Arctic sea ice

- More frequent and longer droughts
- Increased frequency of heavy precipitation events, flooding and landslides
- Increased cyclone intensity
- Rising sea level

There is a complex set of factors influencing atmospheric circulation, cyclone activity and other phenomena impacting the climate that are not well understood. The changes and consequences will present both near and long-term challenges. Some greenhouse gases are long-lived. Even with no addition of these gases to current levels, it will require hundreds of years to see significant reductions in the level in the atmosphere. Current estimates are that if all of the measures currently recommended to reduce emissions from human activity are implemented, the predicted temperature rise will vary from a minimum of 2°C to as much as 7°C by the end of the 21st century. A rise of more than 2°C is likely to have serious consequences for the human habitat. Current projections explicitly exclude feedback cycles, such as those involving the release of methane and nitrous oxide which have the potential to further accelerate surface warming.

Consequences

The long-term trends in the release of the variety of greenhouse gases into the atmosphere are complex and controversial. Further, the prospects for significantly changing those trends are equally complex and controversial but are not central to the purpose of this report.

Climate change is likely to have the greatest impact on security through its indirect effects on conflict and vulnerability. Many developing countries are unable to provide basic services and improvements, much less cope with repeated, sudden onset shocks and accumulating, slow onset stresses. These effects span the spectrum from the basic necessities of livelihood to social conflict, including protests, strikes, riots, inter-communal violence, and conflict between nations. Climate change is more likely to be an exacerbating factor for failure to meet basic human needs and for social conflict, rather than the root cause. Climate change is already intensifying environmental and resource problems that communities are facing. In recent decades, social conflict has been particularly prevalent in Africa. According to the Climate Change and African Political Stability (CCAPS) program at the Strauss

Center, during 2000-2008, over twenty thousand deaths were recorded in Africa during violent, politically destabilizing episodes outside of insurgencies and civil wars.[1] The Department of Defense and the combatant commanders recognize these issues, and to varying degrees, have established initiatives hat incorporate environmental security and disaster preparedness into their security cooperation programs.

The single greatest direct driver of impact on the human habitat is water—too much or too little. Water and water management are key factors to food, health, energy, and economic development. Regional variability in rainfall is an underlying cause, but there is a range of exacerbating factors that are both cause and effect, e.g., population migration, agricultural methods that are no longer sustainable, lack of sanitation and the effect on health and productivity. Population increases demand an increase in agricultural productivity. At the same time, the combination of climate change, unsuitable agriculture practices, poor management of fisheries, and lack of development and management of water resources, particularly in Africa, are obstacles to the needed progress. Over the past half century, the renewable water resource per capita in Africa has decreased by a factor of three.

Water management is essential to sustaining populations. Energy and water are also essential to economic progress. However, a lack of economic resources is a formidable obstacle to water management. For example, in Africa, 95 percent of agriculture is rain-fed with little or no capability for storing or transporting water to deal with the variability in rainfall. Systems to store and transport water are a feature of wealthier nations. Further, given that river basins encompass multiple nations, the mechanisms to manage water across a viable area can, in themselves, cause conflict and population migration.

Roles of the National Security Community

Climate change has the potential for significant impacts on all three of the basic elements important to national and international security- defense, diplomacy, and economics. Dealing with these impacts by mitigating the effects on populations and adapting to change will demand the attention of a broad spectrum of agencies in the national government. This will include the Department of Defense, in support of lead US government agencies. While there will be direct effects on the United States from aspects of climate change to include sea level rise and dramatic changes in weather patterns, the most immediate effects with the highest potential for instability will come from the most vulnerable regions of the world where the United States obtains vital fuel and strategic mineral imports and combats terrorism. To deal with these issues, the Department of Defense, the Department of State, and the United States Agency for International Development (USAID) will be particularly challenged.

The United States has neither the resources nor the influence for an open-ended commitment to addressing the world's challenges related to the consequences of climate change. The United States does have a vital interest in promoting stability in areas of strategic interest. A key to success will be extensive advanced planning and collaboration with others most influenced by the impacts. Near-term work to deal with the immediate basic needs of populations will demand a multiagency and multinational response. Lasting progress will come from longer-term adaptation to climate change. Adaptation will inevitably include more effective water management, population migration, changes in agricultural practices, and approaches to dealing with hydrometeorological disasters resulting from extreme changes in weather patterns. The effectiveness of adaptation will have significant national and international security implications.

The most extreme effects will be in areas with limited expertise and financial resources. The United States will need to collaborate with the political, economic, and military leadership in these regions to develop the needed expertise in civil engineering, hydrology, energy, agriculture, land use, and infrastructure planning. The long-term stability of these regions will depend on progress in all of these activities, even with no further climate change.

The United States has a long history of effective response to disasters with both the capability and commitment to respond quickly and effectively, often including a significant role for the military, and in particular the National Guard, in support of civil authorities. The US also has a history of successfully dealing with the need to adapt with a long-term focus. Examples include one hundred years of the Agricultural Extension Program that led a farm revolution and the Public Health Service that fostered changes in sanitation and food and water processing that virtually eradicated a set of debilitating childhood diseases. Similarly, the Army Corps of Engineers and the Bureau of Reclamation have dealt effectively with water transportation needs and flood control. This kind of sustained attention will be required to deal with near-term needs and adaptation to climate change.

There is existing structure and activity across the whole of government that can provide much needed expertise. A management and cooperation structure is needed to focus increased attention to assisting vulnerable regions in adapting to climate change. Examples of existing activities include:

- United States Global Change Research Program
- Climate Change Adaptation Interagency Task Force
- Department of State Special Envoy for Climate Change
- Department of State Regional Environmental Hub Program
- DOD Environmental International Cooperation Program
- DOD Minerva Initiative: The Climate Change and African Political Stability program
- Central Intelligence Agency Center for Climate Change
- Department of Agriculture Foreign Agricultural Service
- Environmental Protection Agency Climate Program Office
- National Oceanic and Atmospheric Administration Tsunami Resilient Communities concept work
- Geographic combatant commanders theater security cooperation plans, supported by environmental security engagement activities
- United States Geological Survey
- Military support for civil authority activities with various partnerships with USAID, the US Geological Society, Environmental Protection Agency, and the Department of the Interior
- United States Pacific Command Center of Excellence for Disaster Management and Humanitarian Assistance
- US Bureau of Reclamation
- Department of Defense Strategic Environmental Research and Development Program
- Department of Defense, Office of Naval Research Multidisciplinary University Research Initiative on socioeconomic-political driven migratory response of populations affected by rising sea levels

As suggested earlier, while these activities are important, their potential will be significantly enhanced with a structure and process for coordination to more effectively leverage the efforts to address global problems. These efforts need to include better insights into what other countries and international organizations are doing.

Role of the Department of Defense

The Department of Defense will inevitably be a part of approaches to adapt and respond to climate changes in both the United States and in key areas of the globe. Building regional capabilities and alliances to create climate change resilience will be an important contribution to regional stability. To be effective, DOD activities will need to be part of a comprehensive multi-department effort and in coordination with international efforts. DOD will play an important role in dealing with the potential for armed conflict driven by climate-driven population migration. There are existing examples of the potential for conflict generated by drought induced changes in farming and grazing practices in the Darfur region of Sudan, the margins of the Sahel, and southern Africa. The most effective influence of climate-related security issues will come from attention well before the situation deteriorates to the conflict stage.

The Department of Defense has demonstrated capabilities to respond to natural disasters. Much of this experience is applicable to dealing with the near-term effects of climate change. Still, there is a major difference. The traditional objective of disaster relief is a return, as quickly and as practical, to the condition of life as it existed before the disaster.

In the case of climate change, there will be no return to the preexisting life's condition. Instead, near-term solutions need to be on the path to adaptation. DOD experience and capabilities include sustained attention to large-scale, long-term challenges. The Department is frequently called on to serve that purpose. DOD routinely conducts contingency planning across a wide spectrum and executes its plans in cooperation with other departments.

The objective in Africa and elsewhere should continue to be sustainable political stability where civilian governments, supported by defense cooperation among militaries capable of supporting civil authority, promote resilience to the effects of climate change. The bulk of effort and influence must come from the affected region.

Still, the Department can play an important role by providing climate change data and warning, and can help formulate programs to assist foreign militaries to understand the effects of climate change on their force structure, installations, and their country's security situation; and training countries on how to build their capacity to provide effective mitigation and adaptation in support of civil authority.

Summary of Recommendations

Recommendations on the Climate Information and Modeling Needs

The President's Office of Science and Technology Policy should expand on the Roundtable for Climate Information Services to:

- Define requirements for information systems, catalog existing resources relevant to those requirements, identify gaps, and produce a conceptual roadmap for addressing those gaps.
- Identify obstacles to sustained availability of climate information with international scope.
- Define an operational framework for sustained translation of climate data records and other geophysical information into societal benefit metrics.
- Identify approaches and mechanisms for providing sustained, timely, and actionable synthesis assessments focused on developing regions and locales beyond the current US focus, including options for growing in-country capacity.

The Administrator of the National Oceanic and Atmospheric Administration should:

- Work with the National Aeronautics and Space Administration to conduct a renewed study of options for increasing the availability of low-cost, high-reliability launch vehicles for civil science satellites critical for climate observations.
- Establish a mechanism for frequent reassessment (annual vs. decadal) of observational needs responsive to changing scientific understanding and impacts due to failures or funding, including an evaluation of impacts of such developments to the operational needs of the DOD.

The President's Office of Science and Technology Policy should work with DOD, Department of State, and USAID to identify priorities for operational (distinct from research) climate data in priority regions.

Recommendations on Roles of the National Security Community

The Director of National Intelligence should:

- Establish, within an appropriate agency of the Intelligence Community, an intelligence group to concentrate on the effects of climate change on political and economic developments and their implications for US national security.
  - An important focus of this effort should be to project human security changes that could develop into national security issues.
  - This group should make extensive use of open sources, seek to cooperate with other domestic and international intelligence efforts, and report most of its products broadly within government and nongovernmental communities.
- The intelligence group should commission the Central Intelligence Agency's (CIA) Center for Climate Change and Security to produce an assessment of regional climate change hotspots that threaten human security and governmental legitimacy and exacerbate existing tensions. They should use this assessment as a confidence-building measure to promote communication between antagonistic peoples or states. This document should be the basis for interagency cooperation at the strategic and regional levels.

The President's National Security Advisor, in conjunction with the Council on Environmental Quality, should establish an interagency working group to develop:

- Coordinated climate change policies and actions across US government entities.
- A whole of government approach on regional climate change adaptation with a focus on promoting climate change resilience and maintaining regional stability.

The President's National Security Advisor should continue to emphasize strategic interagency documents, such as the guidance to the combatant commanders which details the link between climate change effects and the underlying conditions that terrorists seek to exploit and should direct relevant organizations to consider this relationship in developing their regional plans.

The Deputy Secretary of State and the Deputy Secretary of Defense should:

- Follow the example of the successful foreign military training assistance program to fashion education and training programs in the fields most relevant to adapting to climate change, e.g., hydrology, civil engineering, construction, agriculture, biology, and public health.

- Make conflict avoidance a priority in foreign assistance (including security assistance and foreign military sales), development, and defense concept development and planning.
- Develop a strategic communication message that links water and food security and increased storm intensity to regional stability and US national security.

Recommendations on the Role of the Department of Defense

The Deputy Secretary of Defense should:

- Establish a DOD-wide coordinating policy board for climate change impacts on national security. This board's functions should include:
  - A coordinating role on climate change information from the strategic and operational perspective. This would include assessing implications for the force structure, deployment options, etc.
  - Compiling and assessing climate change effects information across the geographic combatant commands to identify implications for regional stability and the development of global and regional foreign military assistance programs.
  - DOD's interagency representative for climate change adaptation matters.
  - Serving as the focal point for information, web-enabled, that can be accessed by other Office of the Secretary of Defense (OSD) offices as well as the Joint Staff, Services, and combatant commands.
- Expand the authorities of the Operational Energy Plans and Programs Office to include operational climate change issues.
- Direct the establishment of a program of climate change adaptation pilot projects in concert with related programs at USAID and other agencies to identify, solicit, and fund pilot projects focused on specific adaptation sectors and locales (e.g., management of regions or villages in Africa and Central Asia). Examples of pilot projects and suggested activities might include, but not limited, to:
  - Embrace and augment the World Climate Research Program Coordinated Regional Climate Downscaling Experiment (CORDEX) for one of the sub-regions in Africa. Apply

CORDEX in concert with an assessment activity similar to the Prediction of Regional Scenarios and Uncertainties for Defining European Climate Change Risks and Effects (PRUDENCE) project.

- Extend the observational, modeling, and synthesis assessment capabilities applied today in the United States in the Upper Colorado River Basin to a priority water resource district in Africa, perhaps linked with the Nile Basin initiative.
- Apply coastal hot spot pilot projects focused on offering local-scale risk assessment and planning for integrated sea level and storm impacts on the coupled water-energy-waste resources and physical infrastructures for megacities such as Lagos, Karachi, and Dahka.
- Engage the United States Global Change Research Program (USGCRP) international research programs, DOD commands and their in-country security partners, and international aid agencies such as USAID in identifying opportunities to share climate change-related information and bringing more visibility into stakeholders' activities.
- Focus on near-term, achievable, and measurable goals to develop and demonstrate end-to-end threads of core information systems while incrementally building in-country capacity and competence.

Office of the Secretary of Defense, Office of the Under Secretary of Defense for Policy and the Director, Joint Staff should direct development of a DOD strategic roadmap for climate change-related efforts that builds on the framework laid out in the US Navy Climate Change Roadmap to:

- Ensure that the guidance to the combatant commanders, once signed, is considered to be adequate by the Services and combatant commands for translating the broad-level guidance offered in the Quadrennial Defense Review into actionable requirements.
- Direct that combatant command missions include non-combat support to address serious climate change-induced US national security vulnerabilities.

The Deputy Under Secretary of Defense for Installations and Environment should assemble an inventory of critical facilities and infrastructure to include an assessment of vulnerability to climate change effects and the means to adapt.

The Director, Joint Staff should:

- Create a holistic approach to climate change, integrating efforts of its relevant directorates: J2 (Intelligence), J4 (Logistics), J5 (Strategic Plans and Policy), and J8 (Force Structure, Resources, and Assessment Directorate).
- Require that climate change and disaster risk reduction be integrated into training and exercises.

The Secretaries, Chiefs of the Services, and heads of defense agencies should:

- Better integrate climate change and disaster risk reduction considerations into exercises, training, and educational materials.
- Establish metrics focused on risk reduction to minimize the impact of climate change on military and support operations, forces, programs, and facilities.
- Develop guidance to ensure climate change resilience in DOD project designs and construction by incorporating climate change risk into design standards for facilities and installations, with emphasis on the elements related to energy intensive and water intensive uses.

The Secretaries and Chiefs of the Services should:

- Assess the Services' engineering organizations and the cost- benefits of using them in assisting climate change adaptation.
- Utilize military to military engagement opportunities with coalition partners to enhance resilience to climate change impacts and disaster risk reduction capacities. In so doing, they should expand consideration of roles for the National Guard and reserves. (For example, knowledge of traditionally non-military skills needed to respond to climate change threats is often found in the reserves.)
- Examine tasking authorities for domestic and international response to natural disaster or other disaster risk response situations. For example, the National Guard could bring important assets to an international disaster, as it already does in responding to domestic disasters.

United States Northern Command, with support from the Navy and Coast Guard, should identify the assets that will be needed to operate in the Arctic to include communication assets, personnel training, ice breakers, and other equipment.

The geographic combatant commands should:

- Identify early warning indicators for those areas critical to DOD's mission set.
- Incorporate the guidance from the Quadrennial Defense Review and DOD Strategic Guidance on energy, security, and climate change into theater campaign plans.
- Create a demand signal by articulating the need to understand the implications of climate change and resource scarcities in their region to support their campaign plans.
- Include in their theater campaign plans energy, food, water, and disaster risk reduction strategies and plans for reducing vulnerabilities within their respective areas of responsibility.
- Harness more systematically resources beyond the traditional combatant command structure, to include the National Guard, and its State Partnership Program, service engineering units such as the US Army Corps of Engineers and Naval Facilities Command, and OSD-led programs such as the Defense Environmental International Cooperation Program and the Strategic Environmental Research and Development Program.
- Conduct systematic regional or even more localized impact assessments to identify trends and where their resources should be focused. To this end, each should request that the CIA Climate Change and Security Center provide a report on climate change effects and hot spots in their respective areas of responsibility. Programs such as DOD's Climate Change and African Stability Project (Minerva Initiative) could also be utilized in such undertakings.
- Include as a Tier 1 objective enhancing the capacity of host nation militaries and civil response readiness groups to plan for, and respond to, natural disasters (e.g., floods, coastal storm surges, and droughts).
- Integrate into their humanitarian assistance/disaster relief and other exercise plans additional climate change-related aspects. These exercises should include interagency activity.

- Promote the concept of coordinated management of shared natural resources like water.

1. Current Observation, Model, and Climate Projection Capabilities

To be effective, actions by the United States addressing the impacts of climate change that are potentially important to national security require extensive advanced planning and interaction with others' interests in advance of significant effects. This requires the ability to assess potential future environmental effects of climate change and their political, economic, and geographic impacts years or decades before those effects appear. Generating such assessments involves the synthesis of observations of the current and past climate system, numerical models and their predictions and projections, non-climatic information, and expert judgment.

Assessments of past and current climate conditions are largely based on observational data with well-quantified uncertainties. Assessments of future climate variability and change are more complex and the needed understanding of the nature and limitations of assessment capabilities are often not available to decision makers. The words prediction, forecast, and projections are frequently used, often interchangeably, in assessments of potential future climate change impacts following the familiar paradigm of weather forecasts, with an assumption that predictions, forecasts, and projections use the same methods and produce equivalent information. However, there are important differences between these assessment methods. The differences are relevant for understanding the capabilities and limitations of present-day climate information. This chapter begins with an overview of these methods, followed by a review of current climate observation and modeling capabilities.

Climate Prediction, Forecast, and Projection Methods

To start with familiar territory, weather forecasts are currently produced by Numerical Weather Prediction (NWP) systems. These systems are made up of four principal components:

- Wide variety of observations (e.g., balloon soundings, surface measurements, satellite measurements, and ocean buoy measurements)
- Numerical/computer forecast model
- A data assimilation system
- Expert interpretation

Observations provide an initial condition for the forecast model that supports a physically-based weather prediction projected for five to fifteen days. For the next forecast, the data assimilation system is used to produce a new initial condition valid for that time by incorporating the new observations that have become available with the model forecast applicable to that time. The latter provides dynamical constraints and fills in the gaps where observations are inadequate. Finally, the results of this forecast are interpreted by meteorologists who apply expert judgment to the final products released to end users. This is a rigorous procedure continually exercised by operational weather services every day.

A close analogy, sometimes referred to as seasonal climate prediction (e.g., one to six month lead time), uses essentially the same tools, resources, and general methods applied in weather forecasting, albeit on longer timescales. Currently, seasonal climate prediction is primarily limited to features such as the El Niño Southern Oscillation (ENSO) or the Indian and Asian monsoons. In contrast to weather forecasts, the longer timescales of climate require consideration of additional physical factors and processes that do not change appreciably over the course of a weather forecast.

Examples of such factors and processes impacting climate include soil moisture, sea ice, biogeochemistry, and in particular the interaction with the near surface layer of the ocean. Such models need to be more comprehensive and less detailed because of computational limitations. Moreover, the observational needs are far more demanding, requiring measurements of a wider set of physical factors and processes. Despite this, there are currently a number of successful, skillful, operational, and seasonal climate prediction efforts for seasonal forecasting, but these are mostly associated with predicting ENSO impacts on global weather and climate patterns. A key limitation of these seasonal climate predictions is that they are currently designed to forecast the large-scale pattern, general character, and statistics of tropical Pacific sea surface temperatures but not the specific evolution of the weather patterns within the affected area.

Two extensions of the weather and seasonal climate forecasting paradigms are relevant to assessments of longer-term climate change. The first, decadal climate prediction, attempts to start from an observed initial condition of the climate state and then models the evolution of the larger- slower scale climate patterns (e.g., ENSO, meridional overturning circulation in the Atlantic Ocean, potential greenhouse gas-related warming over the decade), followed by quantifying the statistics (but not detailed behavior) of the weather associated with these changes. One aspect of the upcoming Fifth Assessment of the Intergovernmental Panel on Climate Change (IPCC) will employ this method to assess nearer-term impacts (e.g., through 2035).

The second extension is often associated with the term "climate projections." This terminology generally applies to century-scale models, typically through the year 2100. The word projection, rather than the words prediction or forecast, highlights that a climate projection is meant to represent the expected evolution and character of the most general features of climate (e.g., global mean and large-scale temperatures, snow and ice cover, sea level rise). The projection includes the statistics of weather, its extremes (e.g., heat waves, droughts, and hurricanes), and the character and changes in modes of weather and climate variability.

For climate projection, the role of observations shifts from providing initial conditions to supporting model development and model validation by comparing the results of climate simulations to past observed climate over a relevant period. Such information is not yet comprehensive in terms of a variety of physical processes that come into play over long timescales. Climate data records need to be sufficiently long to capture the variability and trends over the course of the projected lead time, which may be decades or a century. Finally, climate projections often include a range of potential future physical actions affecting the emission and removal of climate-forcing greenhouse gases.

With the above clarifications in mind, it is evident that both observations and models of the Earth system form the foundation for climate assessments. Hence, the fidelity and utility of climate assessments are directly related to the capability and limitations of existing observations and models. Given the need to explore a broad range of potential future changes in the environment, coupled with socioeconomic drivers and impacts, Earth system observations and models are needed to quantify and understand the past and present climate and to provide objective and physically-based inferences on future climate and its associated impacts. The model results need to include rigorous quantification of the associated uncertainties.

Climate Observations

Comprehensive observations of the Earth system—the atmosphere, hydrosphere, cryosphere, and biosphere—are foundational to assessing past and present climate conditions and supporting efforts to forecast and project future conditions. Their relevance and importance to projecting future conditions derives mainly from their usefulness in supporting climate model development, validation, and evaluation. Climate models are the only means to obtain objective and physically-based projections of future climate. To be useful in serving that purpose, it is essential that observations are well-designed, calibrated, and maintained over time. Models offer mathematical representations of the Earth system with uncertainties limited by the validity of the underlying assumptions and the spatial, temporal resolutions afforded by the available computational resources. In contrast, observational uncertainties are limited principally by temporal and spatial sampling biases, the systematic errors (e.g., bias) associated with the measurement technology, and the random measurement error (i.e., noise) that can arise from both of the above.

Figure 1-1 presents an example of the differences in fidelity that can occur between observations and model simulations.[2] In this case, the model significantly underestimates the observed information.[3] This is not to suggest that observations are always better than models or sufficient alone. Both are needed. Observational data in the absence of models can be noisy and of limited use without the essential link to physical processes. Models without grounding in observations can be fantasy. Observations complement and support models by facilitating study of key processes towards better models, by offering a benchmark to validate and guide refinement in models, and by integration with models through data assimilation. Data assimilation provides the means to produce retrospective reanalysis utilizing as many observations and model constraints as possible.

Observations of the current Earth system state are conducted from a range of vantage points using different measurement techniques. Observational vantage points include the surface and subsurface of land and oceans (e.g., buoys, drifters and ships of opportunity), airborne platforms (e.g., research and commercial aircraft), and satellites in a variety of orbits. Measurement techniques include sample collection for laboratory analysis, in situ measurements, and remote sensing techniques spanning the electromagnetic spectrum. A subset of these techniques is employed to study proxies for past climatic conditions (e.g., analysis of air and water stored in ice cores).

Many observational systems today are focused on addressing fundamental questions in climate process understanding—for example, the global radiation budget, the hydrologic, carbon, and nitrogen cycles, and other feedback processes. Such research is critical to improving the quality of global climate models but often the question being asked can be addressed with global-scale information using sparse observations, so not necessarily providing the denser sampling and regional or local scale information relevant to decision-making.

Essential Climate Variables (ECVs), listed in Table 1-1, have been defined by the Global Climate Observing System[4] in an effort to prioritize observations and close gaps. Assessment of ECV collection status is an ongoing effort, without clear consensus on the current state of completeness. For example, some assessments show fairly complete ECV data collection but only for global scale assessments. An assessment of ECVs providing higher resolution information needed to support regional assessment and downscaling of climate models offers a less optimistic picture.

For example, observationally derived estimates of global mean sea level, an ECV, are generally considered well established. However, shifts in wind, rain, evaporation, and land ice volume can result in local-scale variations in sea level that can be an order of magnitude larger than the global mean increase projected from thermal expansion of the ocean and land surface melt-water runoff.[5]

Comprehensive assessments of the completeness of climate observations are not readily available. To some extent, this is a moving target given ongoing developments in various development programs and national budgets. However, some insight can be gleaned by considering specific examples.

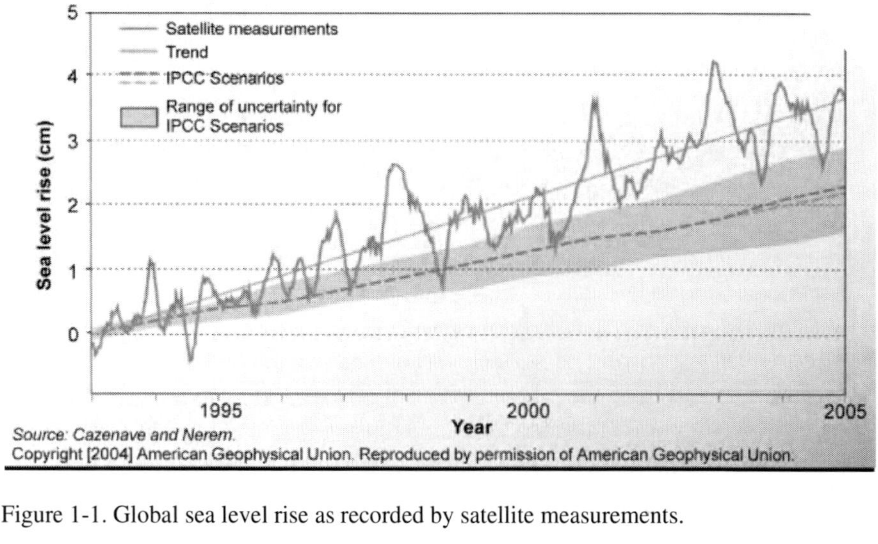

Figure 1-1. Global sea level rise as recorded by satellite measurements.

Table 1-1. Essential Climate Variables

Domain	GCOS Essential Climate Variables	
Atmospheric (over land, sea, and ice)	Surface:	Air temperature, **Wind speed and direction [over the oceans]**, Water vapour, Pressure, **Precipitation**, Surface radiation budget
	Upper air:	**Temperature, Wind speed and direction, Water vapour, Cloud properties, Earth radiation budget (including solar irradiance)**
	Composition:	**Carbon dioxide, Methane, and other long- lived greenhouse gases, Ozone and Aerosol, supported by their precursors.**
Oceanic	Surface:	**Sea-surface temperature, Sea-surface salinity, Sea level, Sea state, Sea ice**, Surface current, **Ocean colour**, Carbon dioxide partial pressure, Ocean acidity, **Phytoplankton**
	Subsurface:	Temperature, **Salinity**, Current, Nutrients, Carbon dioxide partial pressure, Ocean acidity, Oxygen, Tracers
Terrestrial	River discharge, Water use, Groundwater, **Lakes, Snow cover, Glaciers and ice caps, Ice sheets**, Permafrost, **Albedo, Land cover (including vegetation type), Fraction of absorbed photosynthetically active radiation (fAPAR), Leaf area index (LAI), Above-ground biomass, Soil carbon, Fire disturbance, Soil moisture**	

Note: Bold font indicates ECVs primarily provided by satellite observations.
Source: *Committee on Earth Observation Satellites (CEOS) 2011.*

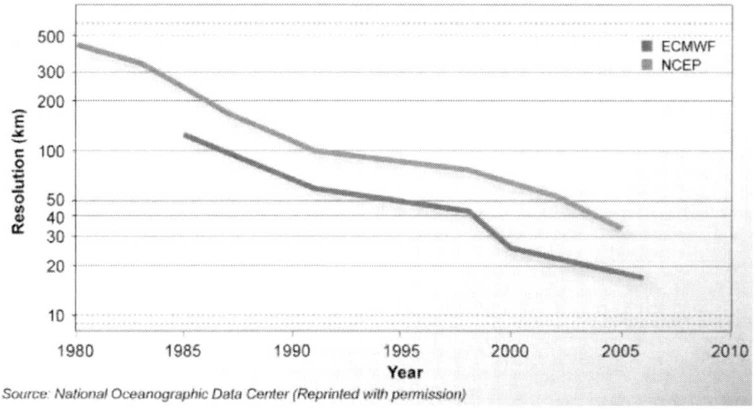

Figure 1-2. Trends in short-term weather (0-14 Days) numerical weather prediction resolution.

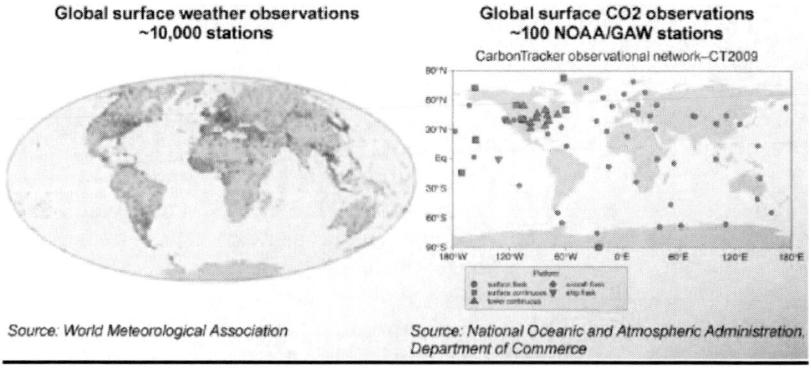

Figure 1-3. Relative density of existing surface weather stations in the United States and global CO_2 observations.

Examining the evolution of Numerical Weather Prediction over the past several decades is instructive. NWP is the foundation of the weather forecasts generated by national weather services. As noted in the opening of this section, NWP uses data assimilation to ingest a variety of in-situ and satellite observations for the purpose of generating the accurate initial conditions for their weather forecasts. Figure 1-2 illustrates a tenfold improvement in spatial resolution in short-term weather forecasting between 1980 and 2005 for the National Centers for Environmental Prediction (NCEP) in the United States and the European Center for Medium-Range Weather Forecasts (ECMWF). This improvement derives mainly from an increase in computational resources and speed which, when combined with improvements in the models themselves and the increasing number and types of observations, has led to a continuing increase in NWP forecast skill over the same period. Over the last decade or more, the above framework of utilizing observations in conjunction with data assimilation methodology and numerical forecast models has air quality and hydrology extensions to weather forecasts. For longer timescales, it has extended to ocean information for the purposes of forecasting El Niño Southern Oscillation and seasonal climate prediction. Most recently, this framework has been extended to consider decadal climate prediction and efforts to employ observations of atmospheric concentrations of carbon dioxide (CO_2), an ECV, to provide potential retrospective analysis and prediction information to climate treaty verification.[6] Figure 1-3 offers a striking comparison between the density of global surface weather stations, an operational system used to support NWP, and surface CO_2 observations from the Global Atmosphere Watch, an exploratory science program, and the

National Oceanic and Atmospheric Administration (NOAA). The orders of magnitude difference in weather versus CO_2 observations and associated funding (Table 1-2) also reflect this stark disparity. The implication for CO_2 is that significant gaps will need to be closed to offer an improvement in capability analogous to that experienced with the evolution of NWP.

Table 1-2. Comparison of total observation assets and annual expenditures (FY09) on the global weather system and global CO_2 observations including satellites in geostationary and low earth (polar) orbits

	Weather	**Surface CO_2**
Surface-based stations	**>10,000 globally**	100 globally
Geo satellites	**>10 (all nations)**	**0**
Leo satellites	**>10 (all nations)**	**1***
Product spatial Resolution (reanalysis and forecast)	**10–40km**	**>2000km**
U.S. annual funding (average, civilian)	**$3,000M** ($1,000M NWS, $2,000M NESDIS)	**$120M** (across 7 agencies, USGCRP)

Note: The asterisk indicates the current capability lacks the required precision and spatial resolution for decision support.

Source: World Meteorological Association.

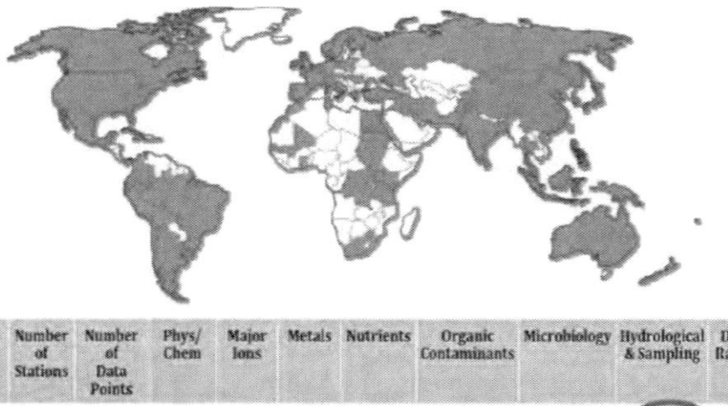

Region	Number of Stations	Number of Data Points	Phys/ Chem	Major Ions	Metals	Nutrients	Organic Contaminants	Microbiology	Hydrological & Sampling	Date Range
Africa	138	2069097	62619	102612	9617	67076	1732	4846	313	1977–2007
Americas	662	417994	188374	237890	307041	220634	594344	19758	12997	1965–2006
Asia	332	641940	214495	151144	82380	117784	8646	36764	12256	1971–2007
Europe	318	823323	249178	134365	182091	146846	23985	40061	66983	1978–2005
Oceana	94	206650	212018	11159	3199	39713	1438	5249	18000	1979–2006
Total	1544	2296814	926684	637170	584328	642053	630145	1006678	110549	1965–2007

Source: UNEP GEMS Water Programme (Printed with permission).

Figure 1-4. Comparison of hydrologic observation sites for UNEP GEMS program.

With regards to surface hydrology observations, Figure 1-4 offers a summary of existing United Nations Environmental Program (UNEP) Global Environment Monitoring System (GEMS) data collection resources by continent. Africa significantly lags other regions in terms of observational density with only 3 percent of hydrologic measurement sites spanning a region representing 20 percent of the Earth's inhabited land surface area.

As of June 2010, the National Aeronautics and Space Administration's (NASA) plan for satellite observations relevant to water resource management includes the following funded programs:[7]

- **Measurements.** The Global Precipitation Mission, with a 2013 scheduled launch, will provide an important advance over the Tropical Rainfall Measuring Mission by providing enhanced geographical coverage, dual frequency radar for precipitation type, and three-hour temporal resolution coming from its constellation of satellites using passive microwave sensors.
- **Soil moisture and freeze-thaw state.** The 2014 launch of the Soil Moisture Active Passive (SMAP) mission will provide global measurements of soil moisture and freeze-thaw state.
- **Inland water height.** The launch of the Surface Water and Ocean Topography (SWOT) mission, planned for launch in 2020, will provide unprecedented insight into the quantity of water in rivers and lakes and higher resolution ocean surface topography.
- **Ground water measurements.** The Gravity Recovery and Climate Change Experiment (GRACE) Follow-on (FO) mission will provide continental (1.000s kms) information on changes in surface snow, ice, water, and stored ground water building upon the approach demonstrated with the GRACE mission.
- **Glacier volume changes.** The Ice, Cloud, and Land Elevation Satellite-2 (ICESat-2) and Deformation, Ecosystem Structure and Dynamics of Ice (DESDynI) lidars can provide information about glacier volumes and their changes over time, especially, but not exclusively, at high latitudes.
- **Water quality in coastal regions.** The launch of the Ocean Ecosystem Spectroradiometer instrument aboard the first platform of the Pre-Aerosol, Clouds, and Ocean Ecosystem mission, planned for launch in 2018, will provide new, high quality information on the biogeochemical properties of coastal waters and their implications for ecosystem and human health.

- **Northern latitude land, lakes, and permafrost.** Change in Arctic features is a broad bellwether of climate change. Coupling of data from the Soil Moisture Active Passive mission, SWOT, DESDynI, and GRACE FO will enable an integrated view of the land portion of the water cycle in this key region.

Similarly, in October 2010, the United States Group on Earth Observations (USGEO)[8] offered several recommendations to avoid near- term gaps aligned with seventeen topical areas.[9] However, subsequent changes in funding impacted multiple elements of the above plans for sustained observations. Additionally, failures of the launch of scientific satellites continue to occur. The decision to launch one-of-a-kind spacecraft and instruments on lower-cost, lower-reliability launch vehicles may be compatible with priorities in a program driven by exploratory science. It does not provide for reliable operational data delivery. There is no indication this trend will be corrected in the near future to maintain a reliable deployment schedule for an operational capability. Operational observations from space today with relevance to climate information are primarily limited to the weather satellites of NOAA/NASA and their defense counterparts in polar and geostationary orbits.[10]

The above discussion on observation is consistent with the following from the 2010 National Research Council study *Advancing the Science of Climate Change*:

> An integrated Earth system analysis capability, or the ability to create an accurate, internally consistent, synthesized description of the evolving Earth system, is a key research need identified both in this report and many previous reports. Perhaps the single greatest roadblock to achieving this capability is the lack of comprehensive, robust, and unbiased long-term global observations of the climate system and other related human and environmental systems.11

Findings on Climate Observation

Today's climate observations and models exist as a loose federation of programs at many government agencies, academia, industry, and nongovernmental organizations (NGOs). While some of these assets are operational systems (e.g., those supporting weather and seasonal climate like the ENSO forecast services), the majority of observational assets and many of the modeling assets today are intended primarily for exploratory science rather than supporting operational, long-term climate assessments.

Many of the observational assets of the civilian scientific community are not robust to failures or data gaps.

Many observations are intended to help improve basic climate process understanding such as closing global-scale energy, carbon, and water budgets but do not necessarily offer the spatio-temporal resolution, completeness or accuracy to support needed improvements in, or validation of, regional climate models, particularly in developing countries.

The US government needs a scientifically robust, sustained, and actionable climate information system that addresses these and other issues. The needs for such a system and key barriers to its establishment are described in Appendix A of this report.

Climate Models

Climate assessments and projections rely on three categories of models:

- Global climate models, sometimes referred to as Earth system models, that represent the biogeophysical processes of the Earth system including atmosphere, hydrosphere, cryosphere, and carbon cycle
- Impact assessment and vulnerability models that address the impact of climate on human resources including infrastructure, health, food, and water availability
- Integrated assessment models which seek to determine the connectivity of human action such as energy, land, and water use with the other two model domains

The three types of models have some degree of overlap. A more complete treatment of these domains for synthesis assessment is presented in Appendix A. This section focuses on the evolving capabilities of Global Climate Models (GCMs).

GCMs vary in complexity and capability but generally consist of coupled atmosphere-ocean general circulation models, with additional components that account for land surface and some cryospheric components (e.g., snow, sea ice). Such models are the backbone of objective, physically-based climate projections.

There are three principal limitations when projecting the combined impacts of climate change in specific regions of the world:

- Uncertainty in the trajectory of future emissions of anthropogenic greenhouse gasses and their airborne fractions in the presence of a changing climate
- Limitations and uncertainties associated with GCMs
- Limitations in deriving predictive information at relatively fine local-to-regional scales (~1 to 100 km) from GCM simulations which are presently limited to representing climate changes on scales larger than about 1000 km

The first limitation convolves the wide range of potential future greenhouse gas stabilization scenarios with the shortcomings in coupled carbon-climate models. In 2001, the Intergovernmental Panel on Climate Change (IPCC) developed a family of future emission scenarios, the Special Report on Emission Scenarios (SRES). The SRES scenarios, illustrated in Figure 1-5, formed the foundation of the range of climate projections associated with the third and fourth IPCC assessments. In the figure, colored lines indicate the six illustrative SRES marker scenarios. The 80th percentile range of subsequent "post-SRES" scenarios comprise the gray shaded areas. Dashed lines show the full range of post-SRES scenarios. Emissions cover CO_2, CH_4, N_20, and F-gases.

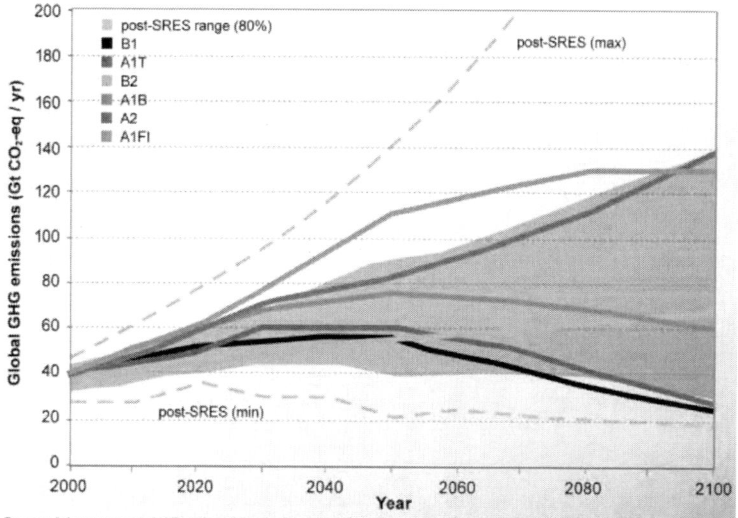

Source: Intergovernmental Panel on Climate Change, 2007: Summary for Policymakers: Synthesis Report. Figure SPM5 of report. (Printed with permission)

Figure 1-5. Global greenhouse gas emissions.

The socioeconomic assumptions behind the SRES scenarios lead to a divergent set of potential outcomes regarding airborne fractions of greenhouse gasses over multiple decades. It should be emphasized that in past IPCC assessments, only a subset of such emission scenarios were fully evaluated by GCMs and subjected to integrated assessment of impacts and vulnerability. To some extent this was due to practical limitations in computational capacity that precluded running full GCMs for all emission scenarios. For example, in the IPCC Fourth Assessment Report (AR4), the most intense scenario A1FI (the FI representing fossil intensive) received partial treatment with Simple Climate Models that attempted to predict the warming that the more complex GCMs would have projected. Given this artificial limitation in the assessment, process readers of AR4 were presented with an impression that a global warming of 4°C by 2100 represented a worst-case scenario; whereas, a closer examination reveals that the likely range of the scenarios is 1.6-6.9°C.

There is an ongoing debate about the likelihood of scenario A1FI. On one hand, recent observations of atmospheric greenhouse concentrations indicate a trend consistent with the A1FI scenario, suggesting this scenario may represent a more likely trajectory. On the other hand, we are still only one decade into the 21st century and some experts continue to assert that A1FI paints a "non-credible" picture of future emissions: that fossil fuel use will peak early this century due to resource limitations. In any event, decision makers should be aware of the uncertainty inherent in predicting future emission trajectories and the broad range of possible outcomes, including those that could lead to two to three times greater warming by 2100 than the 2°C warming scenario that has been the focus of some recent policy discussions and an earlier onset of such warming.

The preceding discussion dealt only with uncertainty in the climate forcing from an emissions perspective in an otherwise static environment. However, the other confounding issue in projecting the climate (radiative) forcing is an incomplete understanding of the biogeophysical processes associated with the carbon cycle and the climate feedbacks on the carbon cycle (with implications on the net atmospheric levels of the two most significant anthropogenic greenhouse gases, CO_2 and CH_4). The airborne fractions of CO_2 and CH_4 undergo significant inter-annual variability due to natural processes such as the El Niño Southern Oscillation. The preponderance of evidence suggests the mean airborne fractions of CO_2 and CH_4 are likely to change in the presence of an evolving climate due to impacts on the carbon sequestration potential of the terrestrial ecosystems and oceans.

While there is disagreement about whether current observations are already showing a trend in increasing airborne fraction (difficult to assess given uncertainties in terrestrial carbon cycle measurements), coupled models such as those contributing to the Coupled Carbon-Climate Cycle Model Intercomparison Project (C4MIP) have shown unanimous agreement that this mechanism is a positive feedback. The airborne fraction is expected to accelerate in a warming climate. Recent coupled carbon-climate models and expert assessment report a best estimate for the A1FI scenario of a 4°C mean warming by 2070-or in the early 2060s if the carbon cycle feedback is stronger than predicted.

The second limitation mentioned above involves the present-day shortcomings and challenges of developing accurate GCMs that form an essential component of our synthesis of the climate system. Equally, if not more important, is their role in providing objective, physically-based climate projections, and thus a backbone of our assessments. Up to, and including, the 2007 IPCC Fourth Assessment, these GCMs accounted for the physical interactions between the oceans, atmosphere, land surface (e.g., soil moisture, land cover type, and influences of vegetation), and elements of the cryosphere (e.g., snow cover, sea ice).

At present, modeling groups are developing their contributions for the next set of coupled model experiments, referred to as the 5th Coupled Model Intercomparison Project (CMIP5)[12] that forms the principal model simulation and projection component for the upcoming IPCC Fifth Assessment Report. For CMIP5, the state-of-the-art in the above-mentioned components continues to evolve and improve, with more realism in the representation of aerosols and the microphysics of clouds and their interactions together, with solar and infrared radiation in the sophistication of the land surface processes, namely vegetation, snow pack, runoff, and soil moisture, of clouds and aerosol, and in the thermodynamic and dynamic representation of sea ice. Moreover, there is capacity for a slight improvement in horizontal resolution, now about 100km for many models when running century-long simulations and projections.

This increased computing capacity, along with additional component understanding and need for better decision support, have necessitated two significant additions to past CMIP model archives and associated science and assessment activities. The first includes explicit carbon cycle coupling within a suite of historical simulations and century-lead projections. For these cases, rather than specify the carbon dioxide concentration, the emissions are specified and the, as yet identified, principal sources and sinks of CO_2, along with transport processes, are simulated to provide prognostic values of CO_2.

These projections allow for more degrees of freedom and will test the understanding of how the land and ocean sources and sinks will respond to the various emission pathways that could be considered.

The second targets more tangible and near-term assessments, with more realism at regional scales, via thirty-year climate projections. These are characterized by using an ensemble of observed initial conditions from the recent past climate (e.g., 1975, 2005) and the associated GCM model predictions from these initial conditions, with the possibility of slightly higher resolution given the shorter integration time. Decadal prediction is still very much in an experimental stage, and the degree of predictability of the natural variability at decadal timescales is still uncertain. In any case, both the century-scale projections (whether carbon-coupled or not) and, in particular, the decadal scale predictions will serve as useful input to regional downscaling efforts to bring more information to decision support issues (e.g., local/regional hydrology, sea level, and extremes).

Given the significant number of global models (~20), each with differences in the manner they model the physical system, along with the multi-member ensemble of simulations and projections expected from each model, and the array of specification for CO_2 (e.g., levels of specified concentration or emission scenarios). These models provide a significant advance in phase space sampling to better quantify scenario, model, stochastic, and initial condition uncertainties. Such information can be utilised from the global models, as well as augmented with downscaled information from a suite of regional models, to provide global to local policy and decision makers a broad sampling of physically-based information. In addition to the above primary new elements to CMIP5, there is a host of science-motivated experiments designed to improve understanding of the physical climate system and provide improved information on climate change and variability attribution.

While the model simulation and projection component of the next IPCC assessment, namely CMIP5, is expected to be an advance over past assessments, there are a number of processes that warrant better understanding and modeling fidelity. Still missing are realistic interactions with soils, ocean chemistry, and deep ocean floor sources and sinks, and processes. Coupled interactions with glaciers and ice sheets have yet to be incorporated, with implications on the fidelity of future sea level rise projections on local and regional scales.

Illustrative of model shortfalls are future projections related to water availability in Africa. The IPCC Fourth Assessment Report[13] provides the latest synopsis of model projections for Africa along with a discussion of model limitations. To summarize, warming in Africa during this century is very likely to be larger than the global annual mean warming throughout the continent, and in all seasons; rainfall is likely to decrease in much of Mediterranean Africa, the northern Sahara, and southern Africa, and increase in East Africa. However, it continues to be unclear how rainfall in the Sahel, the Guinean Coast, and the southern Sahara will evolve.

In addition to these micro- to meso-scale physical issues, on a macro- to global-scale the relative importance of changes in sea surface temperatures and land use change on rainfall is not well understood. Dust and black- carbon are accounted for in many cases but still carry large uncertainties.

The continued and increasing focus on vulnerability and impacts in the upcoming IPCC Fifth Assessment Report, coupled with the new element of decadal-lead predictions (e.g., thirty years from 2005), places greater emphasis on regional downscaling models. While the GCMs are improving their horizontal resolution, the nominal 100 km grids are still much too coarse to represent critical physical details in the landscape (e.g., finer topography, detailed river routing, and vegetation variations) or the interactions of the circulation and hydrology over the landscape. To help address this gap, the GCM output, at the needed temporal sampling (~ <= six hours) and spatial grid specifications, is provided under some CMIP scenarios that allow for dynamical downscaling with higher resolution.

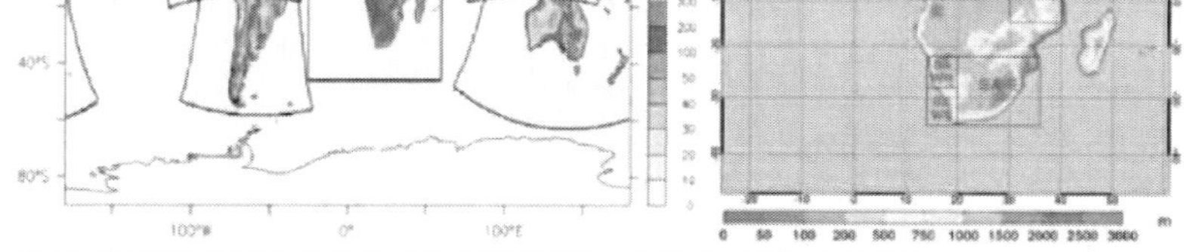

Source: Coordinated Regional Climate Downscaling Experiment (Reprinted with permission)

Figure 1-6. Sub-domains for CORDEX regional climate model downscaling, including Africa.

Regional Climate Models (RCM) activity, as applied to climate change, has proliferated in the last decade or more. In fact, as part of the modeling experimentation for the IPCC Fifth Assessment Report framework, a framework called the Coordinated Regional Climate Downscaling Experiment (CORDEX) has been initiated whereby several specific sub domains (Figure 1-6) have been identified for systematic RCM downscaling.[14] The CORDEX effort builds on two earlier European efforts, Prediction of Regional Scenarios and Uncertainties for Defining European Climate Change Risks and Effects (PRUDENCE) and Common Ensembles Climate Forecast System (ENSEMBLES) and one North American effort, North American Regional Climate Change Assessment Program (NARCCAP) that primarily focused on these respective continents.

In all these cases, more than one RCM is used to form an ensemble of possibilities and account for the uncertainties and individual model biases. RCMs are faced with the same challenges of model shortcomings and evaluations of their fidelity as GCMs with typically the same issues: clouds and convection, aerosol-radiation-cloud interactions, snow and soil processes, etc. In fact, RCMs in some sense are more challenged because of their attempt to represent more processes, particularly those related to impacts.

Similar to the GCM industry, the modelers are taxed simply by the model development efforts, including the numerous processes at a finer, more detailed level, and are not able to always take advantage of new and sophisticated observation resources to evaluate their model and either improve it or quantitatively characterize its uncertainties. Moreover, the process of effectively conducting this sort of experiment (e.g., several RCMs using several GCMs for boundary conditions) is very demanding in terms of computational capacity and workforce. The time, expertise, and resources currently available for these sorts of efforts are not commensurate with the critical role such models will play in future assessments.

Findings on Climate Models

Carbon-cycle models, including their interactions with the water cycle and interactive vegetation models, are in their infancy.

Further development is needed in a number of modeled processes, e.g., cloud processes, soil moisture, interactive snowpack and vegetation processes. The modeling also needs to address how this may influence aerosol sources; sea ice and ice-shelf interactions with the ocean; and, large- scale thermo-haline circulation in association with the details of ocean vertical mixing, convection, and horizontal eddies.

The influences and feedbacks that can arise from melting ice sheet and glacier driven sea level increases, albedo reductions, and ocean circulation modifications are not yet accounted for in the present-day coupled climate modeling framework with significant implications on the accuracy of future sea level projections on regional and local scales.

The inability to provide accurate, regional-scale, long-range precipitation projections is driven by continuing challenges in understanding and representing moist physical processes in the atmosphere and their interactions with aerosols (e.g., biomass burning or dust in the case of Africa) and radiation as well as the limitations of coarse spatial resolutions of climate-scale models.

Despite recent progress towards improving their utility for supporting vulnerability and impact assessments, current climate models are still largely research assets and hampered by limitations in computational and personnel resources.

Increased personnel, education, and funding resources for model development are required to provide the needed more accurate and robust climate projections.

Recommendations on Climate Information System Needs

The President's Office of Science and Technology Policy should expand on the Roundtable for Climate Information Services to:

- Define requirements for information systems, catalog existing resources relevant to those requirements, identify gaps, and produce a conceptual roadmap for addressing those gaps.
- Identify obstacles to sustained availability of climate information with international scope.
- Define an operational framework for sustained translation of climate data records and other geophysical information into societal benefit metrics.
- Identify approaches and mechanisms for providing sustained, timely, and actionable synthesis assessments focused on developing regions and locales beyond the current US focus, including options for growing in-country capacity.

The Administrator of the National Oceanic and Atmospheric Administration should:

- Work with NASA to conduct a renewed study of options for increasing the availability of low-cost, high-reliability launch vehicles for civil science satellites critical for climate observations.
- Establish a mechanism for frequent reassessment (annual vs. decadal) of observational needs responsive to changing scientific understanding and impacts due to failures or funding, including an evaluation of impacts of such developments to the operational needs of the Department of Defense.

The President's Office of Science and Technology Policy should work with the Department of Defense (DOD), Department of State (DOS), and United States Agency for International Development (USAID) to identify priorities for operational (distinct from research) climate data in priority regions.

2. Current Climate Change Situation and Trends

Global Climate Change Trends

This chapter examines trends in global climate change to provide insight into the overall scale and speed of change. This chapter is intentionally rich in detailed observed data. The purpose is to report the availability of credible data on observable trends. This chapter also identifies some specific areas where there is incomplete data or a lack of credible data. All of this data needs to be viewed in the context of the limitations in observed data, predictions, and projections discussed in Chapter 1.

Climate change is not a smooth process. Natural variability that arises from both internal variability, such as the Pacific Decadal Oscillation, as well as external forcing from decadal and longer-term variations in solar activity and from episodic forcing by volcanic eruptions can result in irregular, periodic, and/or singular disruptions in the climate system that act as masking functions to the anthropogenic climate signal.

Dimensions of Current Climate Trends

The Intergovernmental Panel on Climate Change provides a series of reports that document observed climate changes and project future changes. The most recent Fourth Assessment Report was published in 2007.[15] However, continued collection of observational data and the faster than anticipated rate of recent changes have led to several more recent reviews that update key elements of the AR4.[16]

Surface Temperatures Increasing

Between 1906 and 2005, the global average surface air temperature increased by approximately 0.74°C. The warming trend increased toward the end of this period (from 1979 to 2005), to a rate of 0.17 °C/decade.[17] The trend fell in the period 2006-2009, with a global mean temperature rise of 0.24-0.28°C.[18]

NOAA's most recent analysis of global surface temperature change states: "Global temperature is rising as fast in the past decade as in the prior two decades, despite year-to-year fluctuations associated with the El Niño-La Niña cycle of tropical ocean temperature. Record high global 12- month running mean temperature for the period with instrumental data was reached in 2010" (Figure 2-1).[19]

The current estimate of continued temperature increase is 0.2°C/decade.[20] We may be entering a multi-decade period of low solar activity that could give rise to a cooling effect of around 0.2°C over the next two to three decades. At the same time, atmospheric aerosols that have been estimated to reduce warming by 1.1°C have been decreasing since the 1990s, slowly increasing the amount of sunlight reaching Earth's surface.[21]

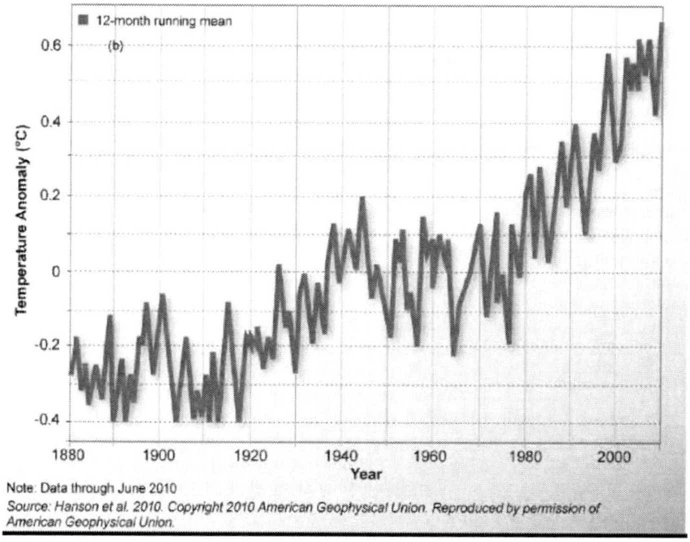

Figure 2-1. Twelve month running mean.

Ocean Temperature Is Increasing

The ocean temperature has a major impact on marine ecosystems, and the associated thermal expansion plays a role in global sea level rise. Increases in sea surface temperature (SST) have been linked to increases in storm activity in the Atlantic. Warming of the Indian Ocean appears to have been a significant factor in the increase of drought in East Africa.[22] A southward shift of the warmest SSTs in the Atlantic has likely contributed to drought in West Africa.[23]

The heat content in the upper 700 meters of the ocean increased over the period 1969-2003 by more than 50 percent.[24] The heat content then reached a plateau in the years 2004-2008 (an earlier reported cooling has been assigned to two systematic biases in the ocean temperature data used) (Figure 2-2).

Figure 2-3 shows changes in global SSTs. In general, measurements indicate that warming is greater in the Northern Hemisphere. The North Sea and the Baltic Sea, in particular, have experienced an unprecedented warming trend since the mid-1980s in all seasons.[25] Temperatures in summers since 1985 have increased at nearly triple the global warming rate and two to five times faster than those in other seasons. In the Southern Hemisphere, each 1°C rise in global average temperature boosted Indian Ocean SSTs by about 1.5°C during the past few decades.[26]

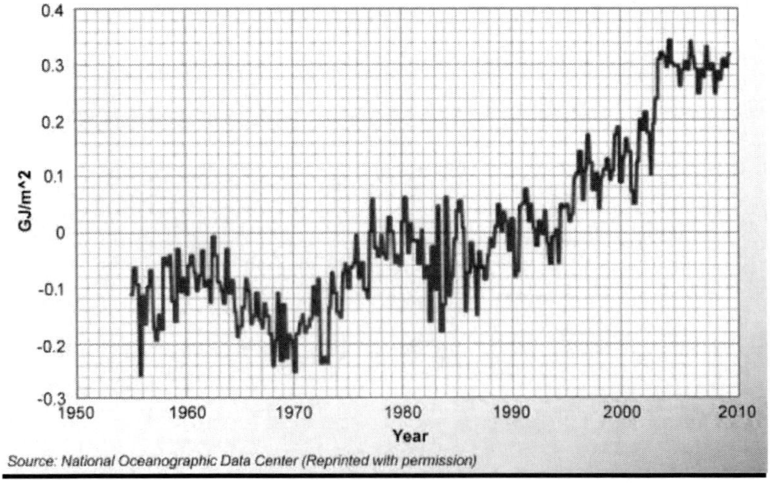

Figure 2-2. Ocean heat content anomalies (0-700 meters), global (January 1955 to March 2010).

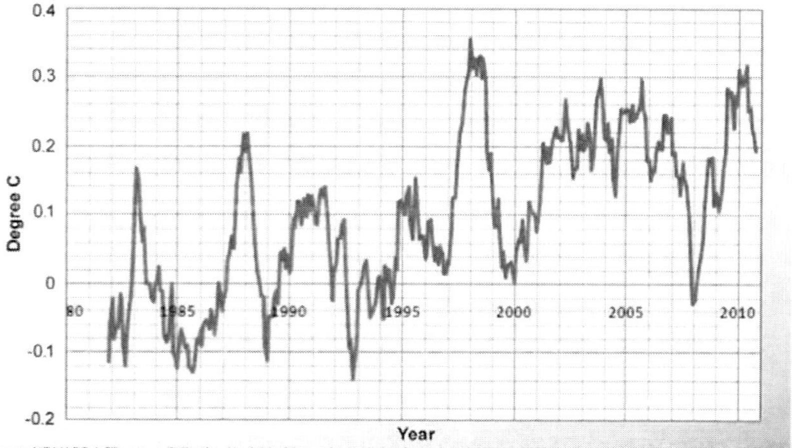

Source: NOMADS: A Climate and Weather Model Archive at the National Oceanic and Atmospheric Administration. (Printed with permission)

Figure 2-3. Global Optimum Interpolation (01) SST anomalies (01.v2) (November 1981 to September 2010)27.

Greenland and Antarctic Ice Sheets Losing Mass

Observational data indicate that melting land ice contributes significantly more to sea level rise than thermal expansion of the oceans. Over the last five years, for example, land ice was responsible for 80 percent of the observed sea level rise over the past five years.[28]

Gravitational measurements from the Gravity Recovery and Climate Experiment satellites indicate that the Greenland ice sheet is losing mass at an accelerating rate.[29] During the most recent eleven summers melting has been greater than the average of the available time series (1973 to 2007). Melting in the summer of 2007 established a new record, which was 60 percent above the previous high in 1998.[30] Between early 2002 and early 2009, the Greenland ice sheet is estimated to have an accumulated ice loss of 230±30 gigaton (Gt)/year.[31]

Satellite observations show that the Antarctic ice sheet as a whole is also losing mass at an accelerating rate. The loss is most significant in the West Antarctic where warming has exceeded 0.1°C/decade over the past fifty years.[32] Ten major ice shelves collapsed in the last decade, and the Wilkins ice bridge failed in spring 2009. Losses along the Bellingshausen and Amundsen seas increased the ice sheet loss by 59 percent in ten years.[33] In the Peninsula, losses increased by 140 percent.[34] In contrast, there were small glacier losses in Wilkes Land and glacier gains at the mouths of the Filchner and Ross ice shelves.[35]

Glaciers and Snow Cover Decreasing

Mountain glaciers serve as important water reservoirs in many countries. Aside from contributing to global sea level rise, glacier melting can result in near-term flooding, including that from the outbreak of glacial lakes, followed by water scarcity. Glaciers and snow cover also reflect solar radiation back into the atmosphere, so decreases in the surface area feed back into increased surface warming.

The average annual melting rate of mountain glaciers has doubled since 2000 compared to the already accelerated melting rates observed in the two decades before. The year 2006 established a new record for annual mass loss of the reference glaciers under long-term observation.[36] The mass lost from glaciers and small ice caps in recent years is estimated to be of a similar magnitude as the total mass loss of the large ice sheets (Figure 2-4).[37]

Arctic Sea Ice Extent Decreasing, Thinning Increasing

The potential for access to new energy resources and shipping routes has made changes in Arctic sea ice and is one of the most studied aspects of climate change. The Arctic sea ice also regulates the exchange of energy between the ocean and atmosphere and between Earth and Space; thus, its loss influences temperature, atmospheric circulation, and weather pattern in several ways.

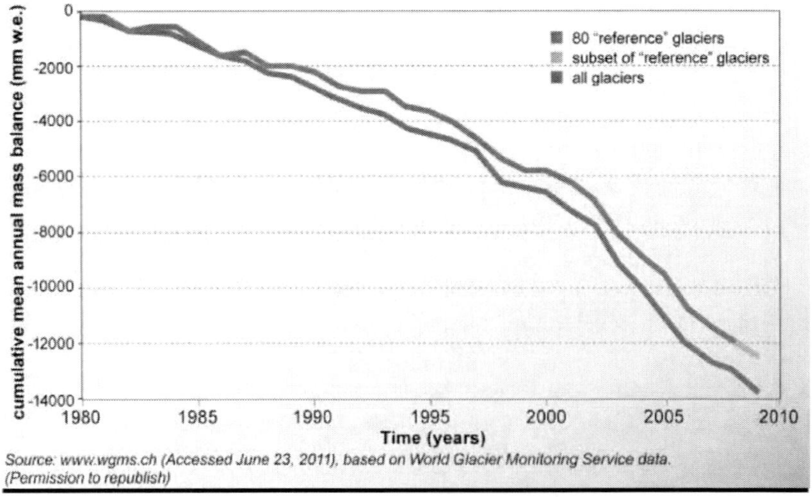

Figure 2-4. Mean cumulative mass balance of all reported glaciers (blue line) and the reference glaciers (red line).

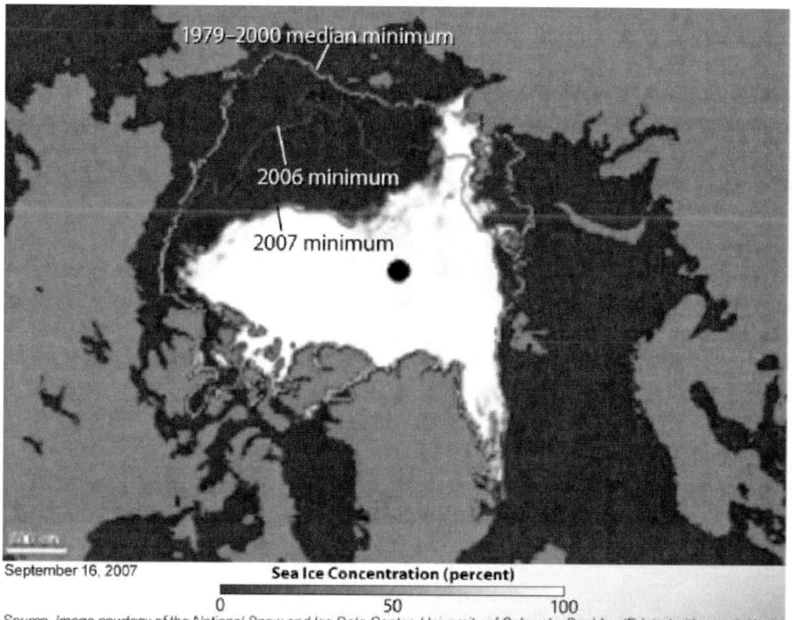

Source: Image courtesy of the National Snow and Ice Data Center, University of Colorado, Boulder. (Printed with permission)

Figure 2-5. Decrease of Arctic ice.

Recent studies predict that the Arctic will be free of summer sea ice sometime between 2030 and 2080, earlier than previously anticipated.[38] Since 1979, the Arctic sea ice extent has been declining at a rate of eleven percent per decade.[39] Ice coverage in summer 2007 reached a record minimum of 4.3 million km^2, a decline in extent by 42 percent relative to conditions in the 1980s.[40] This remains an all-time low. The reduction in the maximum winter extent is smaller, showing a decrease of 2.9 percent per decade.[41] Figure 2-5 shows a time series of the sea ice extent for February and September.

The average thickness of the sea ice has also decreased, increasing the vulnerability to further changes. Thick perennial ice has essentially disappeared, and 58 percent of the multi-year ice now consists of relatively young two- and three-year-old ice, compared to 35 percent in the middle of the 1980s.[42] Submarine sonar measurements covering the central ~38 percent of the Arctic Ocean show an overall average winter ice thickness of 1.9 m in 2008, compared to 3.6 m in 1980.

Figure 2-6 shows the most recent data on annual minimum sea ice extent. Ice extent for September 2010 was the third lowest in the satellite record for the month, behind 2007 (lowest) and 2008 (second lowest). The linear rate of decline of September ice extent over the period 1979- 2010 is now 11.5 percent per decade relative to the 1979 to 2000 average. Changing atmospheric patterns, such as the North Atlantic Oscillation circulation, may lead to a partial short-term recovery of Arctic sea ice depending on the effect of increased warming.

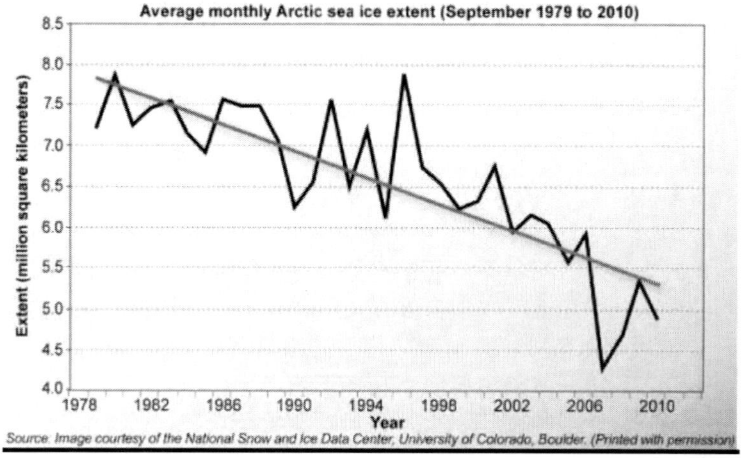

Figure 2-6. Average monthly Arctic sea ice extent, 1979 to 2010; area of seasonally frozen ground is decreasing.

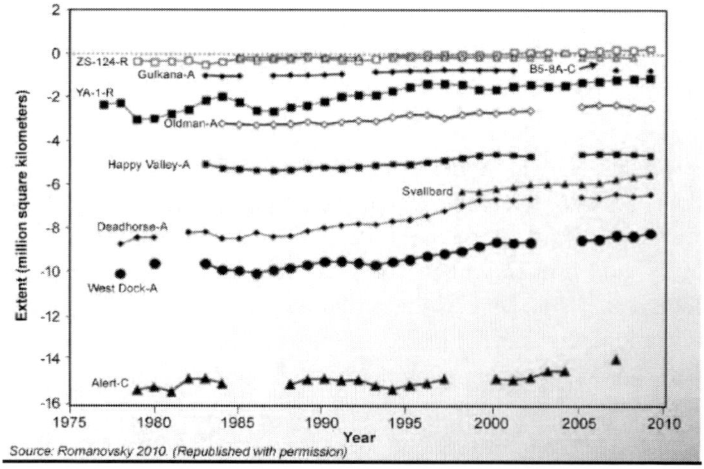

Figure 2-7. Mean annual ground temperatures between 10 and 20 m for boreholes in the circumpolar northern permafrost regions.

Thawing of frozen ground can have major impacts on infrastructure and transportation. Of more significance to the global climate, the process of exposing organic carbon to microbial decomposition results in emissions of carbon dioxide, methane, and nitrous oxide.

This positive feedback cycle is significant because methane is thirty (roughly) times stronger than CO_2 in terms of its short term warming potential. Although not as strong in short term effect, N_2O is important because it is a particularly long-lived greenhouse gas.

Thawing is occurring at a faster rate than anticipated. For example, the maximum area covered by seasonally frozen ground has decreased by about 7 percent (up to 15 percent in spring) in the Northern Hemisphere since 1900.[43] The temperature at the top of the Arctic permafrost layer has increased by up to 3°C since the 1980s.[44] The expansion of Siberian thaw lakes in response to Arctic warming has led to an estimated 58 percent increase in methane emissions, largely from thawing permafrost.

Current estimates of future emissions from these lakes exceed previous estimates by a factor of five or more.[45] Emissions have also been detected in unexpected regions, such as from the submerged wetlands of the East Siberian Arctic Shelf.

Figure 2-7 shows the most recent data on increasing temperatures in monitored permafrost sites in the Northern Hemisphere.[46] Projections suggest that a further decrease in extent of 30 to 50 percent is possible by mid-century.[47]

After a period of slowdown in the growth of atmospheric methane concentrations, measurements since 2007 show a global increase.[48] Estimates of the carbon stored in global permafrost have doubled in the last few years, putting the permafrost carbon stock at an equivalent of twice the atmospheric carbon pool.[49] Potentially, the amount of carbon released during the 21st century could lead to accumulated emissions of around 50-100 Gt Carbon (C) by 2100.[50] Alternative estimates, based on experimental Arctic data, suggest the potential for thawing permafrost to release an additional 38-100 GtC per degree of warming.[51]

Fewer data are available on recent and projected releases of greenhouse gases from methane hydrates that can be found in polar continental sedimentary rocks and oceanic sediment at water depths greater than 300 m. Methane hydrates occur when large amounts of methane are trapped within a crystal structure of water, forming a solid similar to ice. The size of these reservoirs is uncertain.

More Frequent and Longer Droughts

Increased drying in the tropics and subtropics due to higher temperatures and decreased precipitation have increased global dry areas, as indicated in Figure 2-8.[52]

The percentage of global land surface in drought has increased by 3.4 percent per decade in the period 1952-1998.[53] In terms of the Palmer Drought Severity Index, areas in severe and extreme drought have more than doubled since the 1970s, with a large jump in the early 1980s due to an El Niño Southern Oscillation-induced precipitation decrease and a subsequent expansion primarily due to surface warming.[54]

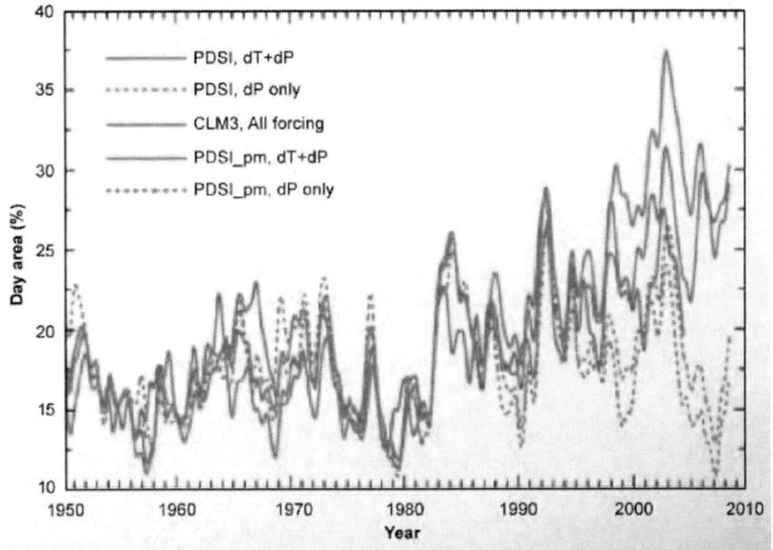

Note: Based on the CLM3-simulated top-1 m soil moisture content (green), and Palmer Drought Severity Index calculated with both observed precipitation and temperature and Thornthwaite (red solid line) 15 and Penman-Monteith (magenta solid line) Precipitation-Evaporation (PE), and with precipitation only (dashed lines). Monthly data were used in the Palmer Drought Severity Index and PE10 calculations with variations on <12-month time scales were filtered out before plotting.

Source: Dai 2010. (Printed with permission)

Figure 2-8. Time series of global dry areas as a percentage of the global (60°S-75°N) land area.

Figure 2-9 shows which regions have experienced the largest change in droughts over the last fifty years and the projections for the next fifty years. These projections indicate how severe the problem is expected to become, particularly across the tropics and subtropics. Based on decreases in soil moisture, the number of short-term (four to six month duration) droughts is expected to double from the mid 20th century to the end of the 21st century. Long-term, more than twelve month duration, droughts are projected to become three times more common.[55] Using a different model and lower emissions scenarios, the proportion of the land surface in extreme drought is predicted to increase from / percent for the present day to 30 percent by the end of the twenty-first century.[56]

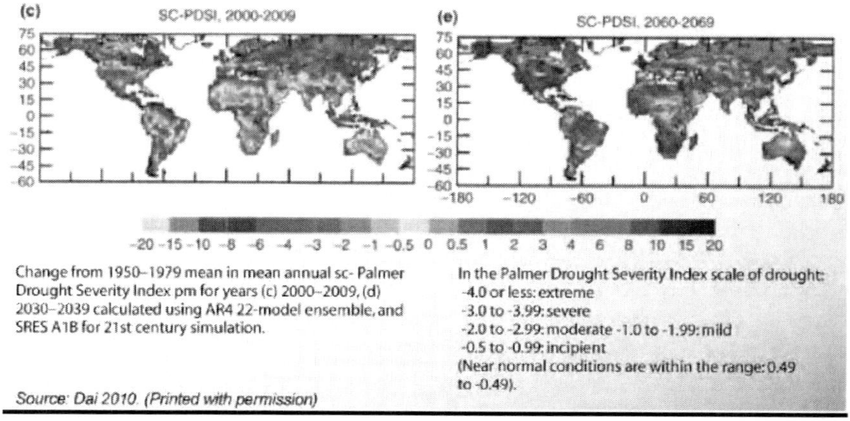

Figure 2-9. Observed and projected changes in drought, in terms of mean annual adjusted Palmer Drought Severity Index (22-GCM ensemble, SRES scenario A1B)[57].

Increasing Frequency of Heavy Precipitation Events, Flooding, and Landslides

Globally, there has been no statistically significant overall increase or decrease in precipitation, although trends have varied widely by region and over time.[58] Figure 2-10 shows global precipitation anomalies.

Long-term trends over the 20th century show a general pattern of precipitation increases at higher northern latitudes and a drying in the tropics and subtropics over land. Significantly increased precipitation has been observed in eastern parts of North and South America, northern Europe, and northern and central Asia. Decreases have occurred in the Sahel, the Mediterranean, southern Africa, and parts of southern Asia. Spatial and temporal variability have increased over the century. Increases in extreme precipitation have been observed even where overall precipitation has decreased,[59] with the amount of rain falling during the heaviest precipitation events increasing by 3-10 percent per degree of warming.[60]

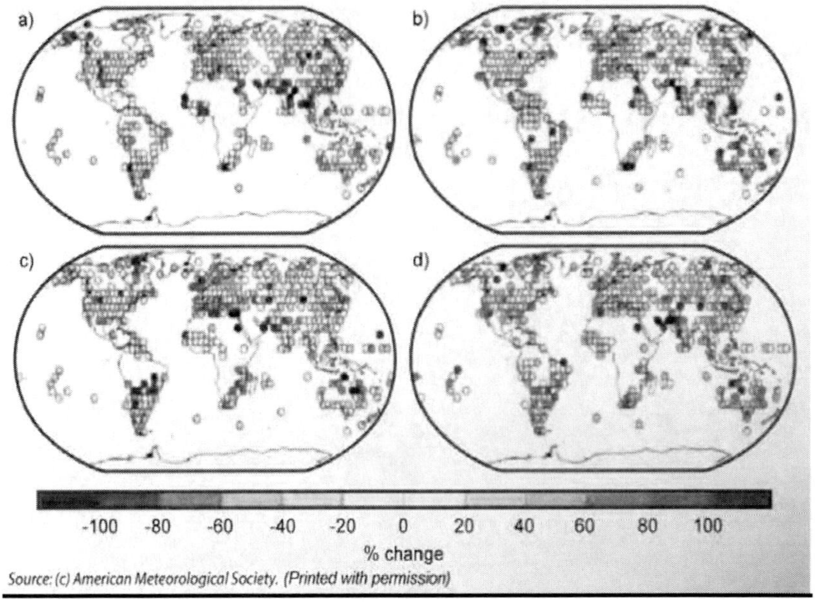

Figure 2-10. Global precipitation anomalies61.

The number of great inland flood catastrophes from 1996-2005 doubled, compared to 1950 and 1980, and has been associated with an increasing frequency of heavy precipitation events.[62] The contributing effects of increasing human impact on the environment and increasing exposure of people and property to flood damage are uncertain.

Projections of changes in the number and intensity of future floods are scarce. One approach is to look at the exceedance of peak volumes in river flow. Figure 2-11 shows how one hundred-year peak volumes of monthly river flow are likely to be exceeded more frequently with a quadrupling of CO_2 in fifteen of sixteen large river basins. In some areas, the current hundred-year flood is projected to occur as frequently as every two to five years.[63]

Landslide (including mudslide) activity at national and regional scales has intensified. As with floods, it is difficult to separate the effects of changes in extreme precipitation from those human impacts on the environment.

The high number of small to medium-scale landslides, which are widespread in many parts of the world, cause high costs to human society, in large part due to the economic, political, cultural, and geographical barriers to effective landslide risk reduction strategies that exist in many developing countries.[65]

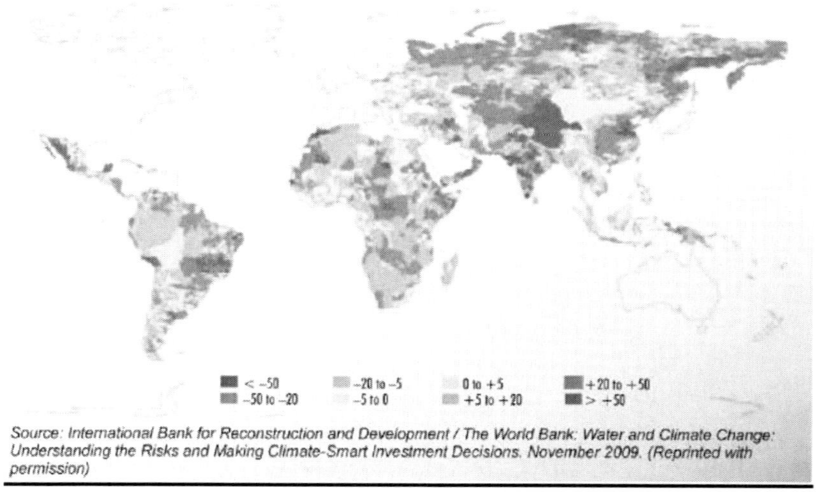

Figure 2-11. Projected percent change in high flows (floods) for 203064.

Extreme Temperatures More Common

Elevated temperatures have both direct and indirect impacts on many important biological systems. Those most commonly discussed are impacts on human health and agricultural growing seasons.

Since 1950, elevated temperatures have increased by between 1°C and 3°C, significantly more than the change in average temperature.[66] Models using moderate emissions scenarios, Special Report on Emissions Scenarios A1B, project that hundred-year return temperature values will exceed a dangerously high level of 50°C in densely populated areas of India, the Middle East, North Africa, the Sahel, Australia, and equatorial and subtropical South America by 2100.[67]

Other recent research combines observation and modeling data to suggest a high probability (>90 percent) that growing season temperatures in the tropics and subtropics by the end of the twenty-first century will exceed the most extreme seasonal temperatures recorded from 1900 to 2006. In temperate regions, the hottest seasons on record will represent the future norm in many locations.

Tropical Cyclone Intensity Increasing

The consequences of increased cyclone intensity are potentially severe, particularly when coupled with sea level rise, and could affect coastal communities around the world.

The destructive power of tropical cyclones (storms, hurricanes and typhoons) has increased since 1970, owing to increases in intensity and duration, as much as 70 percent in the Atlantic and Pacific. The changes in the power destructive index are highly correlated with SSTs in the critical region where tropical cyclones form.[68] Observational data indicate that a 1°C rise in SSTs can increase surface wind speed 5 m/sec, potentially doubling the number of Category 5 cyclones (Figure 2-12).

There is agreement that climate change is likely (>66 percent probability) to increase the occurrence of more *intense* cyclones, although uncertainty remains as to whether this is the result of natural variability in the system or warmer temperatures.

The scientific community similarly agrees that the *number* of cyclones is unlikely to increase. However, a recent study that accounted for the number of North Atlantic cyclones likely to have been missed by the observing system in the pre-satellite era finds a significant increasing trend for 1900- 2006 of ~4.2 storms per century in tropical Atlantic storms (Figure 2-13). The extent to which a minimum in 1910-1930 influenced this trend is uncertain, but this work merits further consideration.[71]

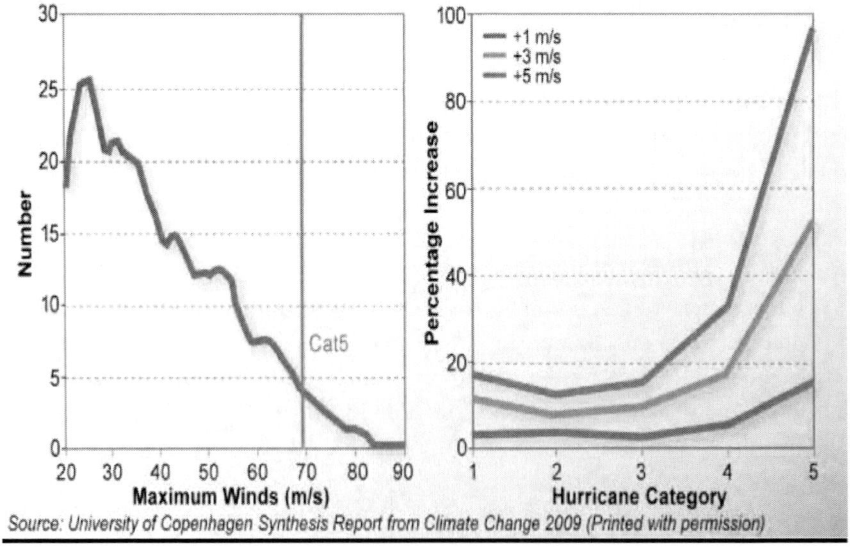

Figure 2-12. North Atlantic tropical cyclones[69].

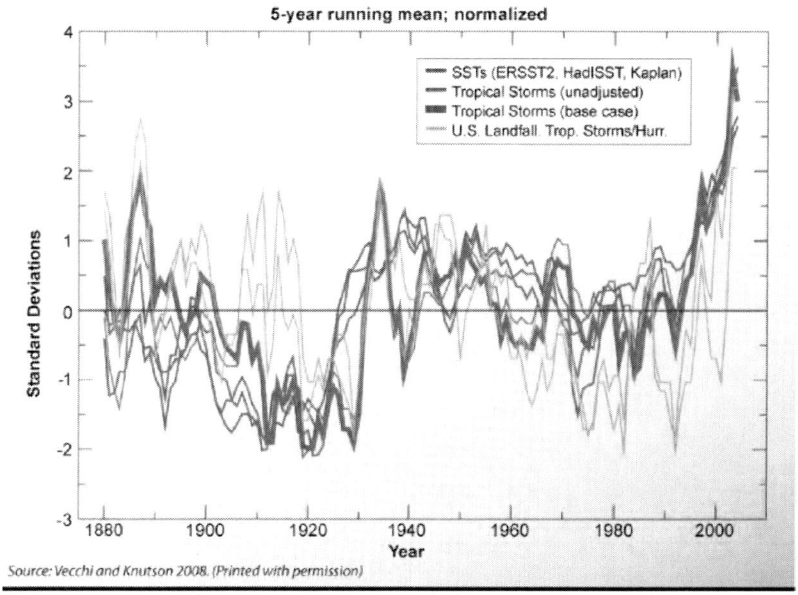

Figure 2-13. Tropical Atlantic SSTs, tropical storms, and US landfall series[70].

Mid-Latitude Wind Patterns and Storm Tracks Shifting Poleward

Mid-latitude westerly winds have increased since 1950, but it remains uncertain whether this is due to warmer temperatures or decadal-scale fluctuations.[72] There is wide agreement on a poleward shift in storm tracks.[73] Any such change leads to changes in regions affected by storm- related strong winds and heavy precipitation, as well as high waves and storm surges.[74]

Sea Level Is Rising

After the last glacial maximum about twenty-thousand years ago, global sea level rose by more than 120 m. The rate of rise slowed down to a steady rate of 0.25 mm/year before the start of the 20th century.[75] Global average sea level rose at an average rate of 1.8±0.5 mm/year during the period 1961-2003, increasing to 3.2 mm/year from 1993 to early 2006.[76] Satellite measurements of sea level rise in more recent years show no major change, with an average rate of increase around 3 mm/year.[77]

Increased understanding of feedback cycles and glacier dynamics has resulted in dramatic increases in projections of future sea level rise, compared to that given in AR4. Figure 2-14 compares the results of these studies to the

AR4 projection.[78] It is increasingly probable that global sea level rise will exceed 1 m by 2100.[79]

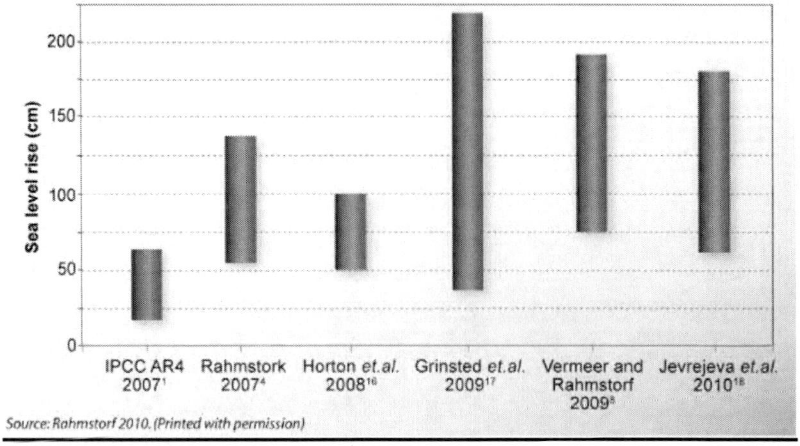

Figure 2-14. Estimates of twenty-first century sea level rise from semi-empirical models as compared to the IPCC AR4.

Crosscutting Points

- Inertia in the climate system means that we are committed to an additional rise in temperature (0.4°C to 0.6°C) comparable to that experienced in the 20th century even if greenhouse gas emissions ceased today.[80] The additional warming is larger when feedback cycles are considered.
- An increase of more than 2°C is commonly cited as having serious impacts on the human habitat. It is expected to exceed this limit in the next few decades. Current best estimates suggest that if all emission reduction efforts registered under the Copenhagen Accord, and subsequently captured in the Cancun Agreements, are fully-delivered, global average temperature is still likely to rise by 3-4°C by 2100.[81]
- Some greenhouse gases are long-lived. Changes in surface temperature, precipitation, and sea level are largely irreversible for more than one thousand years after emissions cease.[82]
- Sea level rise will not stop in 2100. Changes in ocean heat content will continue to affect sea level rise for several centuries at least. Melting and dynamic ice loss in Antarctica and Greenland will also continue for centuries.[83] There are a number of critical thresholds, or

tipping points, at which a relatively small additional forcing induces an abrupt change in the response of some large-scale components of the Earth system. Figure 2-15 provides the global temperatures estimated for tipping points of the most important subset of components.[84] Future policy-relevant tipping elements in the climate system and estimates of the global warming (above present) that could cause their control to reach a critical threshold.

The paleoclimate record shows several periods where abrupt changes occurred. At the end of the Younger Dryas (11,500 years ago), temperature in central Greenland jumped to about 15°C in a single decade.

Findings on Climate Trends

Observational data indicate that changes in the climate system are continuing at a faster rate than reported in the IPCC AR4 (2007). The rate of change in land and sea surface temperatures over the 21st century is expected to at least double that of the 20th century.

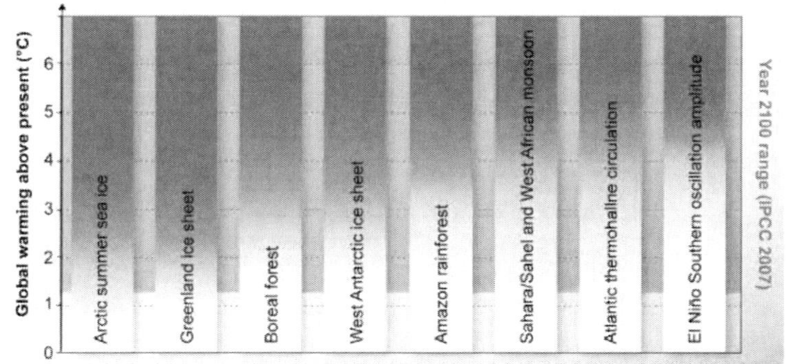

Figure 2-15. Potential policy-relevant tipping elements that could be triggered by global warming this century, with shading indicating their uncertain thresholds.85.

The consequences of continued changes necessitate a risk management approach to climate-related decision-making. Decisions should be based on assuming a global temperature increase of 3-4°C (best case 2°C; worst case 5-7°C) by 2100. This necessitates including the potential for high impact abrupt changes.

Recommendations arising from the climate trends are found in Chapters 4 and 5.

Current Trends and Projections for Africa

Climate changes vary significantly at local levels. The rate of warming in the Arctic, for example, is outpacing that of the remainder of the planet. The remainder of this section provides a snapshot of some of the observed and projected changes for temperature (Figure 2-16) and precipitation in Africa.

Precipitation in the Sahel zone in West Africa shows a general negative trend since 1970, and years of varied positive and negative indexes were more common in the years preceding 1950. The 200 mm/year and 600 mm/year isohyets (contours of equal precipitation) are shifting south with the climate conditions of the Sahel and Sahara also migrating south. The extent of this shift varies across the region but averages about 200 km in West Africa (Figure 2-17). Modeling studies project a significant shortening of the rainy season over much of southern Africa after 2050, an increase in the severity of dry extremes (ten-year driest events) over western Africa that parallels a significant mean decrease in summer precipitation, and an appositive shift of the whole precipitation distribution.[87]

The frequency of anomalously strong rainfall that causes floods has increased in East Africa. Reported floods have increased from an average of less than one event/year in the 1980s, to seven events per year between 2000 and 2006. This is the strongest increase among the full set of hydrometeorological disasters (an increase from an average of less than three events per year in the 1980s, to over seven events per year in the 1990s, and almost ten events per year from 2000 to 2006).[88]

Projections of changes in river flows by 2030 and 2050 suggest increased flooding in East Africa and Central Africa, as shown in Figure 2-18.[89] Another study projects an increase in the intensity of ten-year wettest events that also translates into increasing flood risks for East Africa between 2051 and 2100.[90]

The southward shift of the warmest SSTs in the Atlantic and warming in the Indian Ocean are responsible for a long-term drying trend over Africa and the recent Sahel droughts. Increases in aridity are broadly consistent with trend patterns of independent records of precipitation and stream flow.

The annual variability of rainfall has increased. The chances of drought occurring in parts of the Greater Horn of Africa have doubled from one in six years to one in three years.[91] North African countries now experience five or six years of drought per ten years instead of the one drought every ten years common at the beginning of the century.[92] Dry years have become more frequent in the Volta river basin since the early 1980s and have occurred at shorter intervals. The areal extents to this dryness have also been increasing. The basin recorded at least four moderate dry years covering over 50 percent of the area between 1983 and 2001.[93] The Niger river basin is experiencing a 30% reduction in annual volume.[94] Climate models project increased aridity in the 21st century over most of Africa.[95]

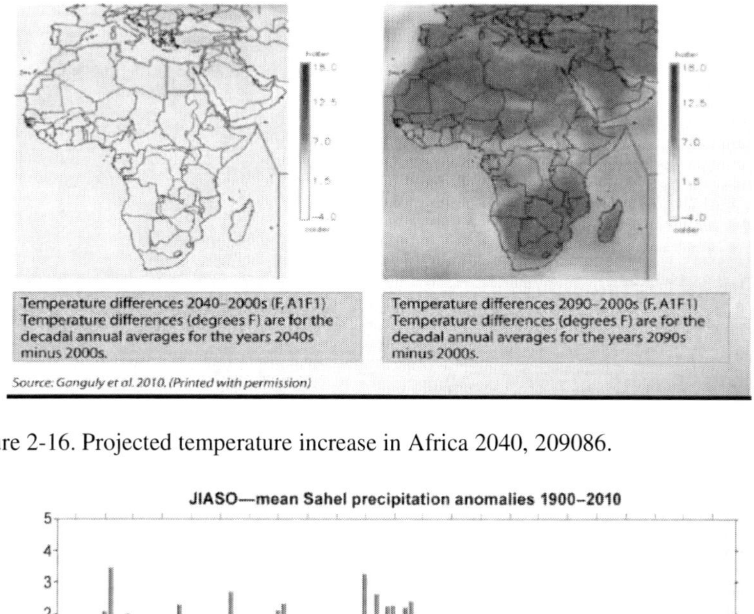

Figure 2-16. Projected temperature increase in Africa 2040, 209086.

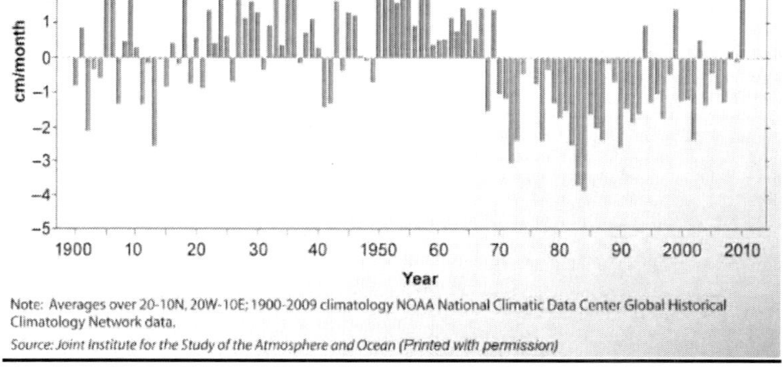

Figure 2-17. Mean precipitation anomalies for Sahel zone West Africa, 1900-2009.

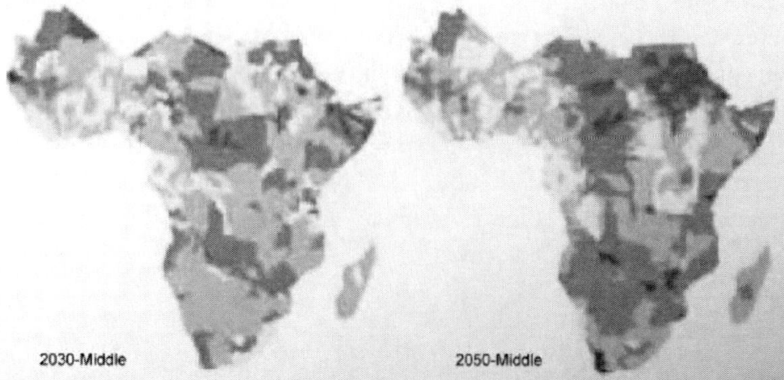

Note: (a) 2030, (b) 2050, (22 GCMs, medium projections across SRES scenarios A2, A1B, and B1).

Source: International Bank for Reconstruction and Development / The World Bank: Publication Water and Climate Change: Understanding the Risks and Making Climate-Smart Investment Decisions. November 2009. (Printed with permission)

Figure 2-18. Projected percent change in high flows (floods) at catchment level compared to a 1961-1990 baseline.

Findings on Climate Trends in Africa

Information on climate change trends in the United States Africa Command (AFRICOM) area of responsibility (AOR) is essential to planning for operations. Better understanding of current and future climate-related impacts can focus military to military training, exercises, and outreach activities on changing needs. In general, information on the regional and local impacts of climate change that is useful for near and mid-term decision-making is inadequate. The range of information deficiencies includes:

- Lack of availability of data on changes in the periodicity of critical events (e.g., return period of floods)
- Lack of assessment of the impact of climate changes on urban environments (Current studies limited to the impact of sea level rise and storm surge on coastal areas)
- Failure to integrate the full set of environmental, socio-cultural, economic, and governmental factors into impact assessments
- Lack of methods and data to support identifying and assessing the cost-benefits of particular adaptation options
- The scenarios and models in use to project continued climate changes lag current scientific understanding.

- Understanding on any one geographical area, type of change, or impact requires synthesis across an increasingly vast array of materials developed by diverse stakeholders.
  - Ongoing DOD activities to address responses to climate changes can also aid partner nations in reducing climate- related risks. US Army Corps of Engineers (USACE) guidance on sea level rise, for example, is applicable to African coastal countries.
  - Lack of visibility into international efforts impedes effective action. Could, for example, RANET (supporting low-tech information dissemination to rural and remote populations) support the African Development Bank's effort on climate and weather forecasting?[96]

3. Potential Consequences of Climate Change

Climate change effects on water availability, food production, health, and local and regional economies can be particularly severe where living conditions, food supplies, health, and governance are fragile. Thus, climate change can be thought of as both an exacerbating factor intensifying existing resource and security problems and as a catalyst creating new environmental and resource challenges (Figure 3-1).[97]

The discussion of consequences in this chapter will focus on Africa, where the capacity to adapt to the effects of climate change is low. Ineffective or inadequate governance will impede near-term responses and adaptation over much of the continent, where potentially helpful institutions are weak. Poverty and disease are widespread, security forces are often unprofessional, and political violence is common. Climate change can impact:

- **Health.** Warmer climate could increase the carriers of vector- borne diseases including malaria and yellow fever. Increased deaths from heat waves, floods, droughts, and malnutrition
- **Food supplies.** Climate change in Africa is likely to adversely affect agricultural productivity. (While the most northern climes on the planet may experience an increase in agricultural productivity, Africa, along with much of South America and South Asia, are projected to see steep declines. The economies of many African nations are still based on subsistence agriculture; well over 50 percent of all Africans

are engaged in subsistence agriculture. More frequent and intensive droughts and floods can lead to food shortages and famines.)

- **Refugees.** Dwindling food and water supplies can trigger migrations leading to conflict at regional, international, and local levels for struggling nations.
- **Coastal regions.** Increased damage from storms and floods and developed regions at risk from rising sea levels

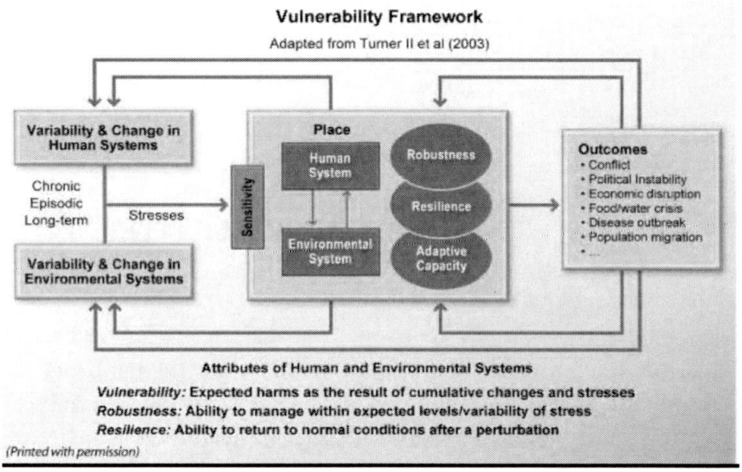

Figure 3-1. Vulnerability Framework.

Table 3-1. Five Climate Threats, Top 12 Most Vulnerable Countries

Drought	Flood	Storm	Coastal Im	Agriculture
Malawi	Bangladesh	Philippines	Micro states	Sudan
Ethiopia	China	Bangladesh	Vietnam	Senegal
Zimbabwe	India	Madagascar	Egypt	Zimbabwe
India	Cambodia	Vietnam	Tunisia	Mali
Mozambique	Mozambique	Moldova	Indonesia	Zambia
Niger	Laos	Mongolia	Mauritania	Morocco
Mauritania	Pakistan	Haiti	China	Niger
Eritrea	Sri Lanka	Samoa	Mexico	India
Sudan	Thailand	Tonga	Myanmar	Malawi
Chad	Vietnam	China	Bangladesh	Algeria
Kenya	Benin	Honduras	Senegal	Ethiopia
Iran	Rwanda	Fiji	Libya	Pakistan

Source: World Bank 2009.

Africa is especially vulnerable to drought and agriculture challenges. Ten of the twelve countries most at risk for each of these two challenges are in Africa (Table 3-1).[98]

The remainder of this chapter focuses on water supply issues.

Impacts of Climate Change on Water

The availability of water underlies all other elements of human security. Changes in precipitation or seasonal delivery patterns can threaten established water supplies supporting agriculture, fisheries, transportation, human consumption, and the natural environment.[99]

Water is a key determinant of land productivity, as well as aquaculture and inland fishery productivity. USAID predicts that global food supplies will need to increase by an estimated 50 percent to meet demand increases by 2030. The Food and Agriculture Organization of the United Nations predicts that water for agriculture needs to increase 60 percent to meet the demand. Clean water supports good health. In Africa, increases in population are outpacing improvements in accessible drinking water and sanitation.[100] New water-related health risks are emerging. The World Health Organization reports that water shortages have driven up the use of wastewater for agricultural production in poor urban and rural communities. More than 10 percent of people worldwide consume foods irrigated by wastewater that can contain chemicals or disease-causing organisms.[101]

Growing populations and rising demands from agricultural, municipal, industrial, and environmental uses increase pressure on water resources. Increased variability and changing patterns of precipitation, melting glaciers, rising sea levels, and increased evapotranspiration are some of the most direct ways in which a warming climate will exacerbate shortages. Societies design themselves around stable climate and resource expectations. Expectations that are unmet or altered significantly on a repeated basis are likely to lead to instability in already weak states.[102] Resolution of rightful use or ownership of water is already a significant issue for local governments around the world.

From 1960 to 2005, the average per capita renewable water resources in northern Africa decreased by more than a factor of three: from almost 3,500 m^3/year to less than 1,000 m^3/year in 2005, with further reduction expected. The African Development Bank estimates that twenty-one countries, nearly half the continent, will experience water stress by 2025, and nine countries will be facing extreme scarcity (Figure 3-2).[103] Declining water levels in many rivers and lakes will likely decrease water quality, exacerbate waterborne diseases, and reduce available hydropower.

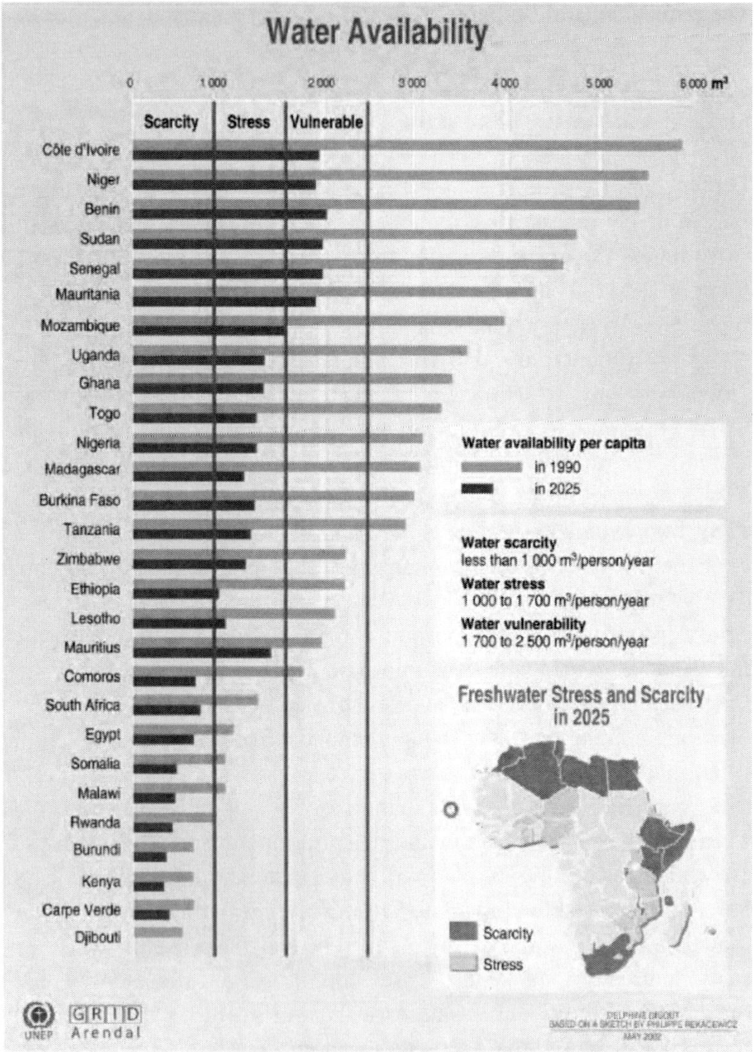

Source: United Nations Economic Commission for Africa (UNECA), Addis Abeba ; Global Environment Outlook 2000 (GEO), UNEP, Earthscan, London, 1999.

Digout, Delphine, based on a sketch by Philippe Rekacewicz; UNEP/GRID-Arendal
http://maps.grida.no/go/graphic/water_availability_in_africa

(Printed with permission)

Figure 3-2. Water availability in Africa.

Unlike most of the world, the primary cause of water scarcity across Africa is a lack of development of available water resources and not physical water scarcity (Figure 3-3). Africa has one-third of the world's major international water basins but can access less than 6 percent of its renewable water resources. The lack of hydrologic data makes it impossible to determine the location and sustainable yield of water in most regions.

Increasing variability in precipitation and higher evapotranspiration rates have forced farmers in many developing countries to rely more heavily on groundwater. As a result, aquifer levels are declining at rates between 1 m/year to 3 m/year, and depleted aquifers may not refill for centuries.

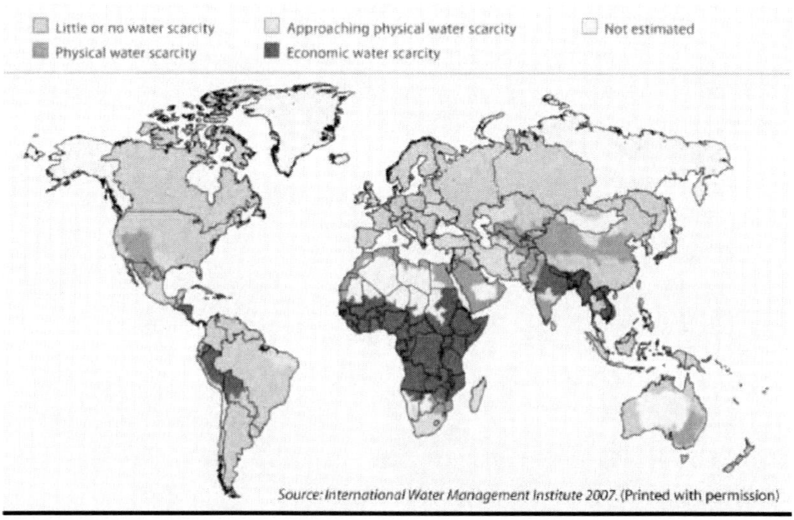

Figure 3-3. Areas of physical and economic water scarcity.

In sub-Saharan Africa more than 95 percent of the farmed land is rain-fed. Current farmers' yields in rain-fed areas are two- to fivefold lower than achievable potential yields, and that current rainwater use efficiency is only 35-45 percent.[104] The lack of storage capacity is one factor that contributes to this heavy reliance on rain-fed agriculture and increases the vulnerability to changes in adequate precipitation. In many African nations, agriculture dominates the economy, which in turn is dependent on variations in precipitation. Figure 3-4 captures the results on gross domestic product (GDP) for Ethiopia and Tanzania.[105]

While water consumption and withdrawal in Africa have been increasing over time, the continent's water resources have been decreasing, mainly as a result of persistent droughts and changing land use patterns. The volume of water estimated to have been lost; for example, from the African landmass during a three-year period ending in approximately 2006, was about 334 km^3, which is as much water as Africans consumed over the same period.[106]

Climate change will intensify existing water issues (shortages, floods, droughts) and effective water management will become ever more critical in developing countries. The many different uses of finite water resources are interdependent. High irrigation demands and polluted drainage flows from agriculture mean less freshwater for drinking or industrial use; contaminated municipal and industrial wastewater pollutes rivers and threatens ecosystems; if water has to be left in a river to protect fisheries and ecosystems, less can be diverted to grow crops. These multiple uses of water complicate water management but offer opportunities for cooperation.

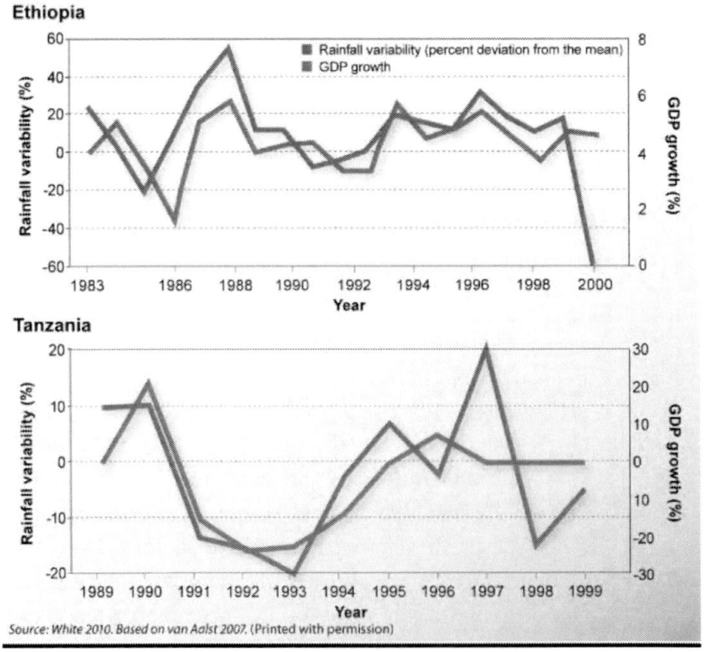

Figure 3-4. Gross Domestic Product growth tracks rainfall variability (inadequate storage & distribution).

DOS and USAID water security efforts are decentralized at the embassy level and do not engage US government technical experts such as hydrologists and engineers in the planning and execution of water projects. The lack of a coordinated US government water security vision and a long term plan, results in a broad and undefined environmental security program at the embassy and combatant command levels, which in turn means that limited resources are not utilized as effectively as possible to achieve US government objectives.

The embassies and combatant commands have a poor understanding of water security issues. Therefore, water security efforts are underfunded by DOS, USAID, and DOD. The Paul Simon Water for the Poor Act[107] is a DOS unfunded earmark, and DOD humanitarian assistance funds are very limited and restrictive. Projects and capacity-building is supported through a myriad of funds such as humanitarian assistance, Humanitarian and Civic Assistance, DOD Environmental International Cooperation Program (DEIC), and 1207/1210 funds.[108] The lack of a dedicated appropriation for water and environmental security projects makes it impossible to develop realistic long-term plans at the embassy and combatant command levels. This leads to difficultly in managing expectations, which ultimately jeopardizes US credibility with partner nations.

Efficient and effective action necessitates a whole of government approach led by DOS/USAID that ensures a single US government approach to water security in Africa and other developing regions.

Recommendations on Water Security

Adopt the DOD/Combatant Command Water Security Program Strategy to institute water security as a core element of DOD strategy that includes:

- Developing a DOS-led interagency team in the combatant command area or responsibility that establishes goals, provides guidance to embassies, and prioritizes efforts to improve water security in regions
- Focusing on areas that meet US government collective interests (e.g., development challenges, potential for conflict, importance of bilateral relations, or existence of forward operating bases)
- Assisting embassies in developing programs that support partner nation efforts to:
    - Locate, quantify, and characterize water resources
    - Identify potential impacts to water resources from climate change

    - Develop long-range plans for sustainable water use and associated land use management for food, agriculture, and livestock
    - Encourage policy and economic practices that promote sustainable water usage
    - Promote the need for clear, and transparent, shared data on water resources
    - Use lessons learned from the United States and other countries' regional water resource management experience (USACE, Environmental Protection Agency, US Geological Survey (USGS), Tennessee Valley Authority, numerous regional commissions, and NGOs) to address co-riparian (shared water resources) equity
- Developing a DOS/USAID lead interagency team at the embassy level to establish a long-term water and environmental security engagement plan for the nation
- Using lessons learned from the United States and other countries' regional water resource management experience (USACE, Environmental Protection Agency, USGS, Tennessee Valley Authority, Bureau of Reclamation, numerous regional commissions, and NGOs) to address co-riparian (shared water resources) equity

Transboundary Issues

Many water troubles are regional problems. Forty percent of the world's population lives on shared basins, which cover more than 50 percent of Earth's landmass. Transboundary water issues are particularly critical in Africa (Figure 3-5)[109] where its major river basins cross many national boundaries (ten for the Niger and Nile rivers; nine for the Congo River), more than those on any other continent. Since one of the primary roles of dams is to manage resource variability and the extremes that can lead to flooding, climate-related changes in precipitation can be expected to increase construction. Drying lands will have a similar effect through increasing demands for irrigation water. However, water projects themselves can be a flashpoint for conflict. Dams, for example, are often beset by problems including forced migration and extensive downstream environmental damage.

The multiple shared interests around water provide a strong economic argument for cooperation and there are trends of cooperative behavior.

However, the likelihood of conflict in river basins rises as the rate of change within the basin exceeds the institutional capacity to absorb that change (as demonstrated by the history of conflicting and cooperative water interactions over the last fifty years). The effects of rapid environmental change or rapid population growth are compounded by major unilateral development projects, a lack of institutional capacity, or generally hostile relations.[110] Changes in precipitation, evapotranspiration, and environment degradation, coupled with increased migration in search of water resources and stable livelihoods, could tip the balance in fragile regions.

Equitable sharing of resources among riparian countries will be increasingly vital in the face of resource unreliability and shortages. Yet the high levels of conflict between African countries such as Ethiopia and Eritrea, or internally, such as in Somalia, prevent governments from establishing meaningful institutional support mechanisms for management of transboundary resources.[111] Existing treaties, especially among contentious partners, may falter with the change in the climatic conditions on which resource management strategies are based. Upstream countries may seek to use climate change as a screen for renegotiating water-sharing agreements to more favorable terms; and, in times of heightened tension over other issues, water infrastructure can become an increasingly attractive weapon of diplomatic pressure, or target in a military confrontation.[112] The shrinkage of Lake Chad illustrates the potential international import of alterations in the geographical distribution of transboundary resources; once shared with Niger and Nigeria, the lake perimeter is now bound by only Chad and Cameroon.

The primary obstacles to effective negotiation are the uncertainties surrounding present and future water uses and water availability. These forestall building trust among partners. Key security unknowns include the likelihood of countries experiencing extreme water scarcity resorting to unilateral actions that generate, for example, retaliatory trade actions, escalating tensions at the international level.[113] Recent years have seen many instances of countries building large-scale dams to meet the needs of their growing populations and economies to the detriment of their neighbors. Potential conflict is now brewing between the neighboring countries of Ethiopia and Kenya as several large-scale hydropower dams are constructed or planned along major stream courses downstream from their headwaters.[114]

A separate issue arises when a country's borders are defined by changing transboundary waters instead of geographical coordinates. Uganda and the Democratic Republic of the Congo, for example, have acted quickly to head off a dispute as changes in the course of the River Semliki over the last fifty years, have transferred as much as 50 km^2 of Congolese territory to Uganda.[115] Border issues like the Migingo Island saga that has pitted Uganda against its eastern neighbor, Kenya, could become more complicated if a further fall in Lake Victoria water levels increases the size of the island above water. Uganda is also looking at border changes occurring along the West Nile and its border with Rwanda in the Katuna wetlands.

These climate-related stresses open up opportunities for conflict-prevention and peace-building. The shared problems are a basis for bringing adversaries to the negotiating table and providing a common language for building cultures of cooperation. History shows that even countries with a history of conflict will come together to negotiate water issues.

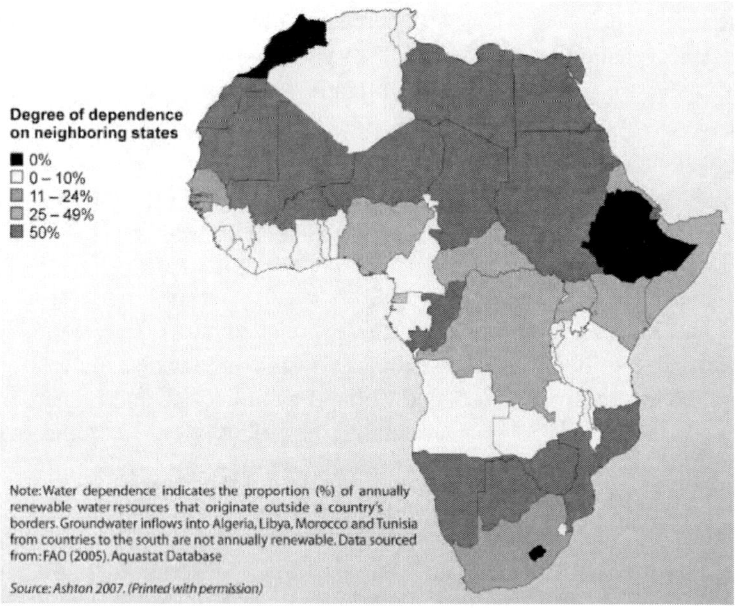

Figure 3-5. Dependence on inflows/water transfers.

Water problems are recognized as "wicked problems" which are further complicated by the need to integrate two very different types of knowledge, work across several socio-political units of analysis simultaneously, and better organize water as a common resource.[116] More fundamentally, most African nations lack data on the extent of their current sustainable surface water, much less data on renewable groundwater resources. Adding increasing climate variability and future climate uncertainties into this mix only compounds the difficulties. Until now, water managers have developed models and tools for a stable climate with known seasonal and decadal cycles. Changes in the variability of precipitation, extreme events, and increases in storm surge in coastal regions raise additional issues for the development of typically long-lived water infrastructure. Tradeoffs between competing uses will become progressively more difficult as demands on those resources increase.

The former President of South Africa and Nobel Peace Prize recipient, Nelson Mandela, stated that: "Security is an all-encompassing condition in which individual citizens live in freedom, peace and safety; participate fully in the process of governance; enjoy the protection of fundamental rights; have access to resources and the basic necessities of life; and inhabit an environment which is not detrimental to their health and well-being."[117]

Consequences for National Security

Climate change is an observable fact. Regardless of cause, recent global temperature rise is outside the range experienced since the end of the last ice age approximately ten thousand years ago. This change has the potential to change many of the delicate balances that affect US national security.

Changes already underway are having, and will have, major consequences for the political, economic and geographic world as we know it. Geographic modifications, driven by climate change, can decide the fate of island territories, even countries, whose highest points are only meters above current sea surge levels. Shifts in river courses and lake levels can result in changing international boundaries.

Climate change's impact on US national security will be determined by its political, economic and geographic manifestations, yet most current climate research seeks to understand and model the climate itself. Modeling climate change physics is like assembling a puzzle. One seeks to find all the pieces. Eventually some ability to forecast rainfall, winds, temperature, sea level, and storm intensity will emerge. The future rate of shift and ultimate magnitude of climate variations now underway are, as yet, impossible to accurately predict, but the direction of these shifts is certain. Earth is warming. It is clear that the shift is a fundamental one that inevitably will alter factors critical to US global interests.

Dealing with the political, economic, and geographic fallout from these physical forces, on the other hand, is a mystery. No amount of data collection will illuminate, with any certainty, actions which result from human emotions, cultures, and personalities.[118] Much work remains in understanding processes for dealing with likely climate-induced conflict types.

In some instances, climate change will serve as a threat multiplier, exacerbating tensions between tribes, ethnic groups, and nations. In other cases, climate change will seem more like Mother Nature's weapon of mass destruction. This natural weapon of mass destruction may also be slow- acting. When rainfall shortage in the Atlas Mountains causes decrease in ground water, salt ocean water can intrude, exacerbating sea level rise related to salinization under North Africa's coastal port cities (e.g., Tunis, Algiers, Rabat, Tripoli, and Casablanca). Fresh water will be displaced, imperiling both potable and agricultural sources.[119] Low-lying ports are vulnerable to damage or disruption resulting from violent storm surges. In time, these currently prosperous centers could become as forsaken as today's Timbuktu; pressure for population displacement to Europe and elsewhere in Africa will grow.

Each of these changes, and many, many more, can strongly affect US national security. Impacts likely will not be directly military. If, and when, military responses are necessary to protect US national interests, responses will be too late to rectify underlying conflicting forces. Climate change will more likely first affect human security, resulting in population and political instability that threatens nonmilitary US interests (access to natural resources, criminal activity and terrorism, economic damage, or political agreements), then escalate to kinetic military conflict.

The climate shifts that produce instability may not occur in the proximity of the disturbance. For instance, the unrest leading to the Egyptian government's downfall in winter 2011, exacerbated by escalating food prices brought on by failure of the Russian wheat crop, could threaten the Egypt-Israeli peace agreement, a matter of US national security. Understanding the political and economic pressures of climate change is a global problem, not just a regional one. Analysis of climate change effects must embrace multiple layers of understanding.

Some Additional Consequences of Continued Changes in Africa

Continuing increase in local mean temperature and extremes, including the frequency and intensity of floods and drought, will compound existing problems that include:

- Reduced fresh water availability
- Reduced agricultural production
- Loss of biodiversity

- Increased food insecurity
- Increased health problems
- Increased migration

The impacts of these consequences can only be understood in their proper context-environmental, socio-cultural, economic, governmental, etc.-that determines vulnerability, resilience, and adaptive capacity.

Developing regions face differing but equally critical climate change-related issues. The one common critical issue is water.'

Agriculture

Climate change may have a graver effect on Africa than any other continent. According to IPCC AR4 projections of a minimum increase in temperature of 2.5°C by 2030, around 600,000 km^2 of cultivable land may be ruined.

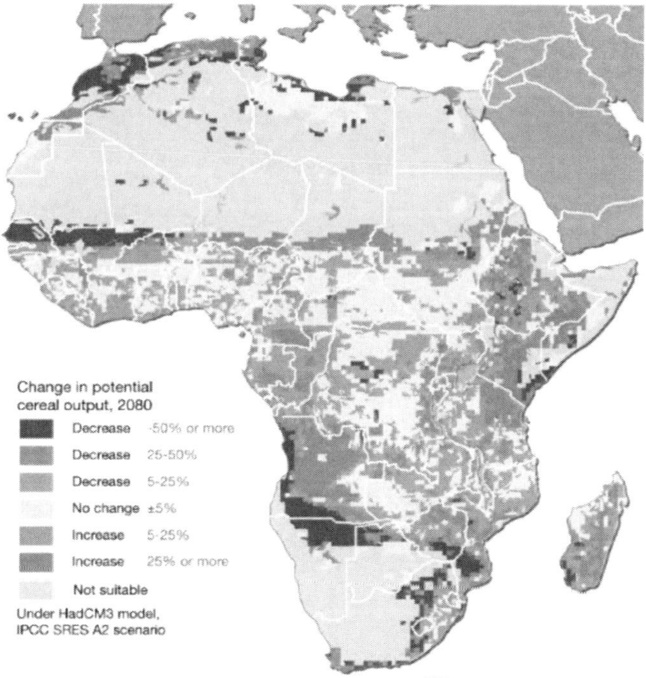

Source: UNEP: The Environmental Food Crisis, 2009.[123] (Printed with permission).

Figure 3-6. Projected climate change impacts for cereal output for 2080.

Several modeling studies have assessed changes in crop yields under projected future changes, with results such as those shown in Figure 3-6.[120] This type of study either excludes the fertilization effect of additional atmospheric carbon or includes assumptions based on laboratory studies. An alternative approach is to project changes in the length of the growing season (Figure 3-7).[121]

Fish are a major source of daily protein for a large percentage of the population. Figure 3-8 shows how changes in ocean temperature and chemistry are projected to change fisheries' catch potential in the oceans around Africa.[122] Changing temperatures and water levels in many of Africa's lakes also affect the productivity of aquaculture and inland fisheries.

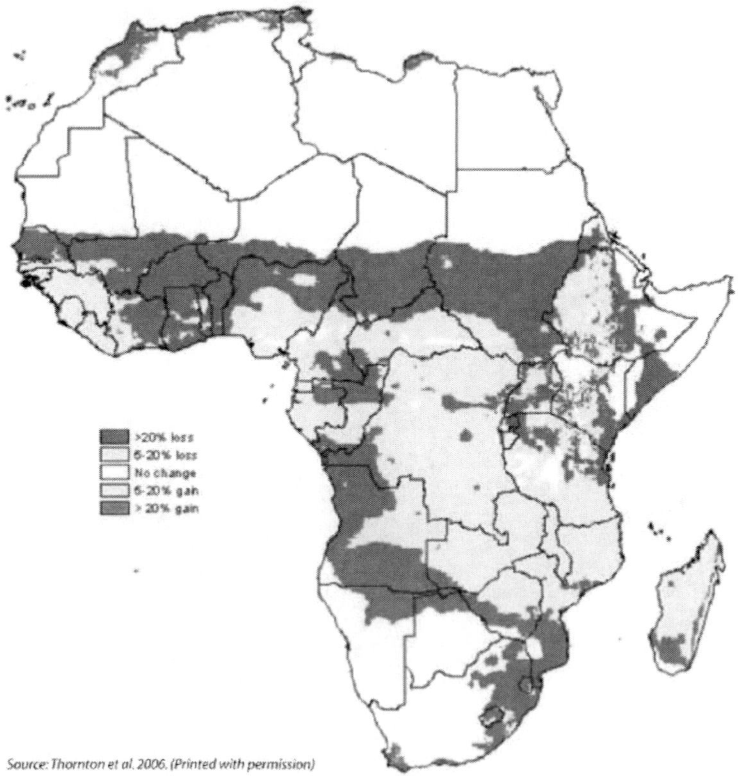

Figure 3-7. Percent changes in length of growing period changes to 2050 (HadCM3D, SRES A1FI).

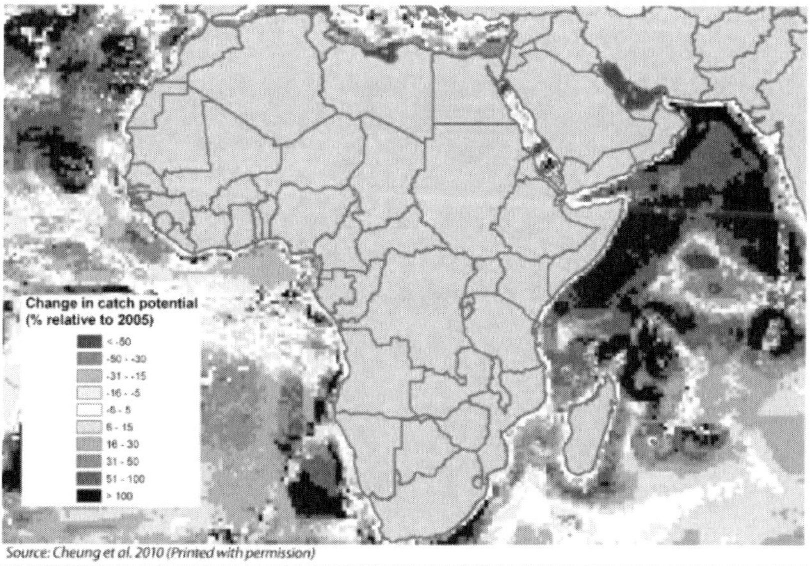

Figure 3-8. Change in fisheries catch potential (10-year average) from 2005 to 2055.

Sea Level Rise

Relative sea level rise is not uniform due to effects that include variation in ocean dynamics, regional variations in thermal expansion, and gravity-elastic influences from land-based ice sheets and glaciers.

A rise in sea level exacerbates storm surge to increase flooding and potential storm-related damage (Figure 3-9). This effect is inversely proportional to the intensity of the storm. An increase in sea level of two feet doubles the storm tide associated with a tropical storm with a normal storm surge of two feet. The storm tide of a severe cyclone with a storm surge of fifteen feet only increases by thirteen percent for the same sea level rise.[124] Even so, a relatively small increase can result in breaching storm defenses. Table 3-2 lists assessed impacts for a projected sea level rise of one meter and a ten percent increase in extreme water (i.e., a one-in-one hundred year storm).[125]

Sea level rise can have several additional consequences. These include coastal erosion, loss of barrier wetlands, saltwater intrusion, pollution from spread of landfill, industrial contamination, and agricultural wastes; and damage to sanitation systems.

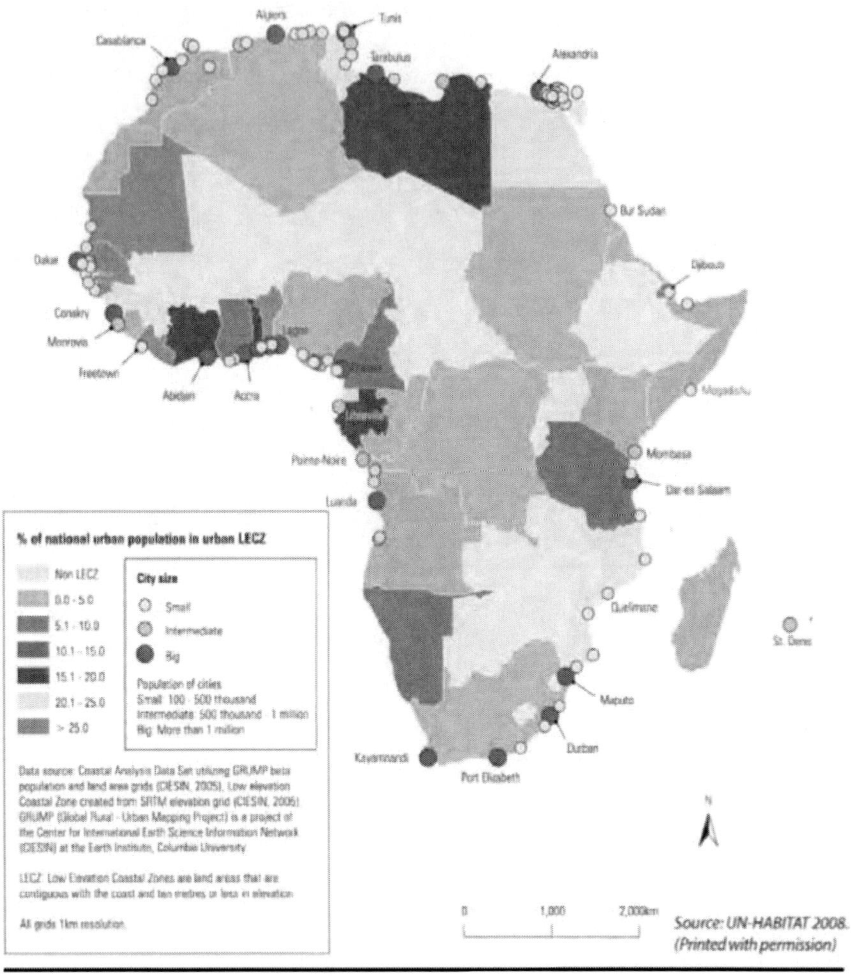

Figure 3-9. Cities at risk due to sea level rise126.

Health-Malaria and Dengue Transmission

The time it takes for the malaria parasite to mature in the mosquito is sensitive to temperature. The parasite takes fifty-six days to mature at 18°C, which is longer than the average mosquito life span but only nineteen days at 22°C and eight days at 30°C.

Figure 3-10 shows the potential change in malaria vectors for 2°C temperature increase and associated precipitation changes.[127]

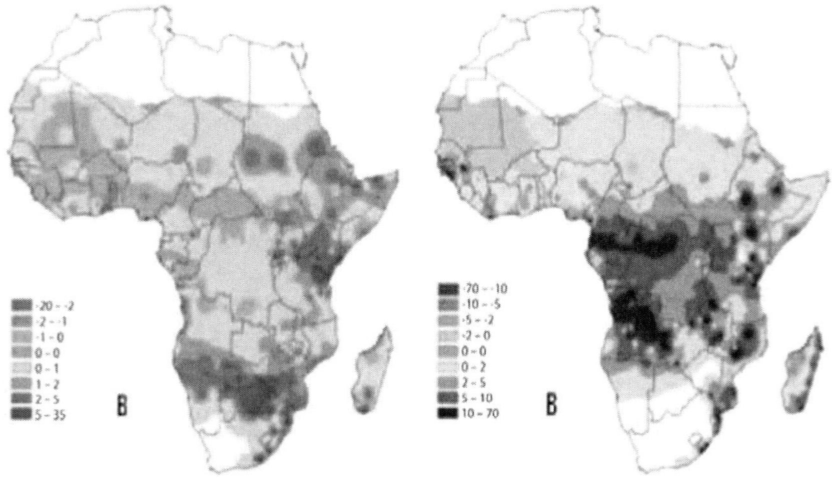

Note: (Left) A. arabiensis, (right) A. gambiae.

Source: Tonnang et al. 2010. (Printed with permission)

Figure 3-10. Change in distribution of the two most prevalent malaria vectors (Anopheles mosquitoes) for a rise of 2°C Africa wide temperature, 10% increase of summer rainfall, and 10% decrease in winter rainfall.

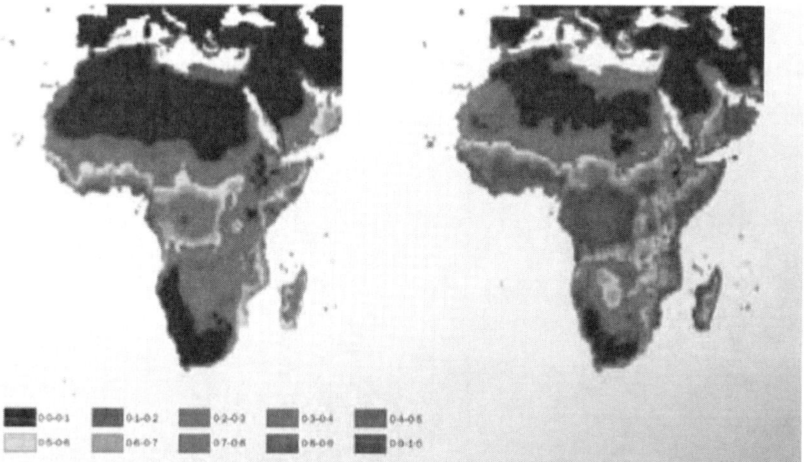

Note: Population at risk of dengue fever: (left) 1990, (right) 2085.

Source: Hales et al. 2002 (Printed with permission)

Figure 3-11. Population at risk of dengue fever.

Table 3-2. Sea level rise (1 m) and storm surge increase (10%) impacts

Country	Land area (km^2)	% Coastal (approx.)	Population impacted	% Coastal (approx.)	GDP impact (USD$ million)	% Coastal domain	Agricultural impact (km^2)	% Coastal (approx.)	Urban extent (km^2)	% Coastal (approx.)	Wetlands (km^2)	% Coastal (approx.)
Mozambique	3,268	41%	380,296	52%	140.73	55%	291	24%	78	55%	1,218	47%
Madagascar	2,312	45%	102,439	43%	27.89	44%	0		36	44%	617	51%
Nigeria	2,264	31%	870,276	25%	407.61	22%	0	100%	94	29%	1,365	39%
Mauritania	1,754	21%	149,576	33%	74.21	35%	0	2%	59	43%	710	33%
Senegal	677	17%	190,690	21%	111.66	21%	29	2%	27	16%	395	22%
Guinea-Bissau	670	36%	61,134	33%	10.01	33%	0		12	34%	278	40%
Cote D'Ivoire	668	30%	315,609	48%	176.27	43%	0		99	53%	162	38%
Gabon	630	26%	34,500	28%	120.95	24%	0		30	30%	253	27%
South Africa	607	43%	48,143	33%	174.30	31%	70	34%	93	48%	132	46%
Somalia	555	28%	33,756	31%	8.90	26%	15	16%	1	25%	94	25%
Sierra Leone	549	29%	39,080	35%	5.69	38%	0		1	37%	451	34%
Namibia	470	60%	957	42%	2.31	37%	0		13	50%	18	82%
Angola	457	29%	72.448	46%	88.54	45%	23	14%	19	46%	129	15%
Eritrea	452	32%	8,238	31%	0.97	29%	0		4	43%	31	32%
Tanzania	426	47%	75,493	50%	34.45	49%	64	22%	15	53%	177	42%
Guinea	420	59%	58,967	44%	37.99	40%	0		8	33%	193	62%

Country	Land area (km^2)	% Coastal (approx.)	Population impacted	% Coastal (approx.)	GDP impact (USD$ million)	% Coastal domain	Agricultural impact (km^2)	% Coastal (approx.)	Urban extent (km^2)	% Coastal (approx.)	Wetlands (km^2)	% Coastal (approx.)
Ghana	400	39%	137,206	49%	45.04	51%	0	67%	35	48%	268	48%
Sudan	370	50%	18,762	49%	10.77	48%	0	0%	7	50%	107	59%
Kenya	274	42%	27,453	40%	10.12	32%	40	22%	9	39%	177	53%
Liberia	269	27%	88,535	45%	16.77	41%	0		15	43%	44	46%
Benin	260	20%	221,029	39%	107.35	47%	0	0%	44	44%	164	21%
Cameroon	172	40%	57,124	35%	44.53	32%	0		14	40%	111	43%
Togo	95	34%	147,274	54%	48.20	54%	1	50%	28	60%	52	27%
Djibouti	82	38%	28,559	60%	22.87	49%	0		5	60%	7	19%
Congo	65	15%	10,361	22%	13.14	22%	0		3	21%	20	11%
DRC	51	17%	1,812	8%	0.17	12%	0		9	32%	21	23%
Gambia	39	4%	47,233	40%	18.54	47%	0	0%	8	24%	21	4%
Equatorial Guinea	22	17%	892	38%	6.32	42%	0	50%	1	53%	4	8%
Sao Tome & Principe	2	44%	1,053	24%	0.30	20%	0	33%	1	30%	0	

Source: Based on Dasgupta et al. 2009, Table 3;International Bank for Reconstruction and Developmen /The World Bank: Sea Level Rise and Storm Surges: A Comparative Analysis of Impacts in Developing Countries, *Public Research Working Paper 4901, 2009.*

The transmission of dengue fever is sensitive to humidity. Figure 3-11 shows how climate change is expected to change the area of land with a climate suitable for dengue fever transmission and, assuming no other contributing factors change, the proportion of the human population that would be at risk.[128]

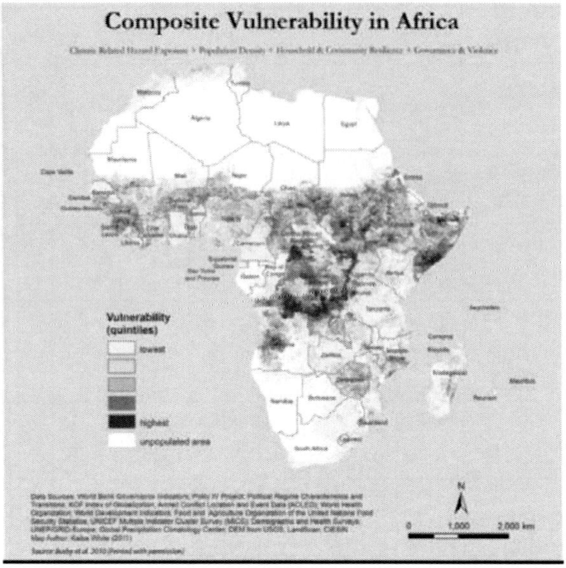

Figure 3-12. Composite vulnerability in Africa.

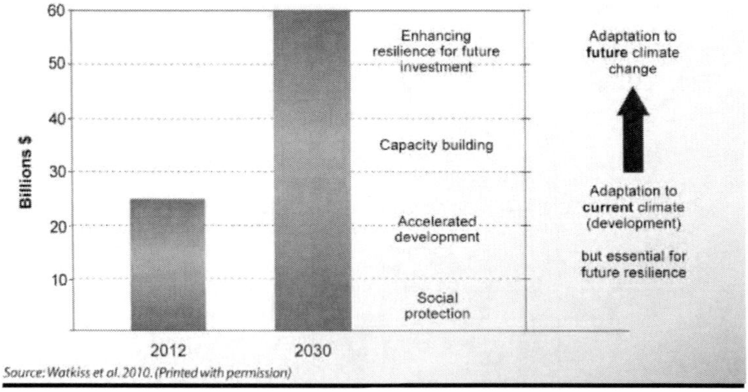

Figure 3-13. Potential costs of adaptation to current and future climate in Africa.

Vulnerability

Funded under the DOD's Minerva Initiative, the Climate Change and African Political Stability program maps vulnerability to climate change as shown in Figure 3-12.[129]

Adaptation

The UNEP-funded AdaptCost Africa project has estimated the potential adaptation costs in Africa. While there is high uncertainty, the integrated assessment models indicate that the central economic costs of climate change for Africa could be equivalent to 1.5-3.0 percent of GDP each year by 2030, as shown in Figure 3-13.[130]

Findings on Consequences

Climate change may have the greatest impact on security through its indirect effects on conflict and vulnerability. Many developing countries are unable to provide basic services and improvements much less cope with repeated sudden-onset shocks and accumulating slow-onset stresses.

Climatic factors affect conflict through influence on food prices, migration patterns, government revenues, competition for water, grazing land, and other resources. The capacity of the international NGO community to fill critical gaps on a much larger scale is at best uncertain.

CCAPS indicates that social conflict-protests, strikes, riots, inter-communal violence, and other unrest have a strong correlation to rainfall. Precipitation in Africa may become more variable bringing more extreme wet and dry years; and, social conflict is more common in extremely wet and dry years than in years of normal rainfall.

Social conflict has been more prevalent than armed conflict in Africa in recent decades, resulting in more than twenty thousand deaths since 2000, many of which came during violent, politically destabilizing episodes.

Climate change is an exacerbating factor. It can be a contributing factor to conflict, but the underlying political, economic, social, and cultural contexts probably will have a more direct effect. Grievances develop, mobilizations occur, and violence erupts more easily where there are weak, corrupt, fragile, or failed governments. Climate change could intensify environmental or resource problems that communities are already facing by intensifying grievances, overwhelming coping capacities, and possibly spurring migration and displacement. Climate change could also create new environmental problems that lead to instability.

DOD and the combatant commands recognize these conditions. AFRICOM and United States Southern Command (SOUTHCOM) initiatives incorporate environmental security, the role of climate change, and disaster preparedness into their security cooperation programs.

Addressing Climate Change Effects on African Populations

As noted in Chapter 2 of this report, climate change effects in Africa destabilize fragile states by overloading the political systems and eroding governmental legitimacy. Many of these states determine continued US success in achieving national security objectives, such as securing strategic mineral and fuel resource access; maintaining freedom of navigation covering choke points and sea lines of communication; combating terrorism; maintaining geopolitical influence; promoting democracy; and, establishing strong market economies. Thus, climate change by definition affects US national security interests. Moreover, DOD has the potential to promote US national security interests with confidence-building measures and by promoting multilateral cooperation and communication. There is compelling evidence of the importance and the potential of DOD's measures. Before the droughts began in Sudan in the 1980s, herders from the north and farmers from the south coexisted and shared the arable land. When drought became a permanent condition, farmers began to fence in their lands to protect increasingly fragile crops. The resulting tensions multiplied existing differences in religion, culture, and ethnicity and underpinned the conflict in Darfur. Similar conflicts occurred across the margins of the Sahel, to include northern Nigeria, and drought has promoted tensions between Zimbabwe and South Africa over forced migration. The questions, then, are these: How are those effects manifest? How should the problem be addressed? What should be the national security community's role in managing it? To manage the problem, one should begin with a clear end state, develop strategic concepts for achieving it, and specify the resources necessary to succeed.

End State

A stable Africa, where civilian governments, supported by defense cooperation among capable militaries, maintain resilience to the effects of climate change, and in which intra and interstate conflict over resources is not a threat to security and stability; and, where preparedness for reducing the risk and responding to increased and natural disasters, is a priority.

Findings on Addressing Climate Change Effects on African Populations

Environmental security and resource issues have been associated with all phases of the conflict cycle; climate change is already implicated in some African conflicts. Climate change increases the porosity of border areas and ungoverned spaces, especially in areas of pastoral landscape. Water security is a driver of conflict, an avenue for peace building, and a critical element of economic growth in Africa. Many African governments lack the capacity to adapt to the effects of climate change. Regional militaries have many of the technical, health, engineering, and manpower capabilities necessary for civil authority's efforts to create climate change resilience. The successful cooperation between the SOUTHCOM-DOS Environmental Hubs on climate change adaptation and environmental security could easily be replicated between AFRICOM and the three DOS African Hubs. This cooperation would make efforts to build climate change adaptation capacity and resilience attainable. Private sector corporations, such as the Coca-Cola Company, and nongovernmental organizations, such as The Nature Conservancy and World Wildlife Fund, have established and respected programs in Africa addressing the climate-related issues such as clean water and food security that could inform AFRICOM efforts to build regional capacity. Because of its widespread occurrence across the continent, climate change adaptation is a viable issue for multilateral cooperation and a confidence building measure for intrastate and interstate conflict. USAID's regional and bilateral missions could harness AFRICOM engagement and capacity building programs to support US sustainable development and climate change initiatives. Recommendations arising from the consequences of climate change are found in chapter 4 and 5.

4. Roles of the National Security Community

This chapter examines the roles of the national security community in addressing climate change issues. Climate change is increasingly recognized as having a multiplier effect for existing tensions and regional instabilities. It places additional stress on the state political system, complicating the ability of governments to meet the demands placed on the system by a suffering population and by reducing system resilience. This can lead to a loss of legitimacy, internal conflict, state failure, population migration, and the growth of extremist ideology. Climate change threatens US national security interests at the regional levels. This study does not address means to halt or reverse global climate change. Lack of knowledge of complex climate mechanisms makes any such suggestion speculative at best. However, the United States has understanding, resources, and skills that can blunt some effects of climate change.[131,132] The United States, however, has neither the knowledge nor the resources needed to produce widespread amelioration. US resources must be focused on the most serious US national risks. The United States must also reach out internationally to spread the burden of adaptation to climate change.

Strategic Concepts for Addressing the Challenges

Strategic concepts include:

- Whole of government cooperation with meaningful support from DOD
- Cooperation with allies, NGOs, and those of common purpose
- Military to military engagement and capacity building

The current US approach to national security interests related to climate change is based on three elements of national power: defense, diplomacy, and economy. While other US agencies can make valuable contributions to this effort, the Department of Defense, the Department of State, and the US Agency for International Development, referred to here as "the 3 Ds," will determine the success or failure of US efforts to reduce the risks associated with climate change in areas of the world of greatest relevance to US national security interests. Success in building cooperation on climate change adaptation will depend heavily on the efforts of these three organizations, supported and guided by:

- An intelligence community analysis of where US national security interests are most likely to be affected by regional changes in climate
- A realistic assessment of the individual capabilities of these organizations to affect climate change adaptation capacity building in the corresponding regions be conducted
- A synchronized plan that capitalizes on the strengths of each organization be created and implemented

Meeting the physical impacts of climate change like changes in water, temperature, sea level, and storm violence are mostly the provinces of civil engineering, hydrology, energy, agriculture and land use, and infrastructure planning. The bulk of resources to address local problems must come from affected regions. However, many of these areas are impoverished and lack even basic technical knowledge. To help address this shortfall, the United States has the capability to build and educate local and regional cadres of engineers, hydrologists, planners, agricultural and fishery specialists, and medical personnel to support local resilience to climate shifts. A major training effort, similar to past US foreign military training assistance programs, could be mounted to educate field personnel in hydrological, civil engineering, construction, agricultural, biological, medical, and criminal justice critical skills in those countries where climate change can be addressed through technical means. The United States should also lead an effort to call on other nations to assist with similar training and education capabilities.

The United States and its developed world allies should be prepared to support, both technically and financially, civil projects that increase resilience and provide ways to adapt to forecasted climate change effects. As in the case of foreign military sales, such support could serve to boost world markets for construction and agricultural equipment, seeds, and medical supplies.

The conventional view of national security is based on conflict and economic interests. Threats to a state are often seen in terms of territory, migration, access to resources (energy, water, food, and materials) or markets. Climate change threats are much less focused. Human security lies at the heart of the climate change threat to US national security. Conventional national security threats arise from human security origins.[133] Dealing with climate change requires a human security-based strategy to prevent emergence of national security challenges.

As a nation, the United States has a reputation for being better at addressing acute crises than at sustaining efforts to remedy chronic problems before they become dire. History suggests the United States is able to do both effectively. The Agricultural Extension Program, in its first one hundred years, led a farm revolution that developed new technology, escalated productivity, battled the Dust Bowl, and dramatically reduced the number of people required to feed the nation. The Public Health Service fostered changes in sanitation and food and water processing that virtually eradicated common childhood epidemic diseases. The US Army Corps of Engineers, since the early /800s, and the Bureau of Reclamation, since the /890s, have altered the US riverine transportation system, enabled growth in the water-starved West, and reduced the impact of seasonal flooding. Mounting each of these efforts required enduring commitment to make changes in the country's interest. Countering climate change's adverse effects on national security will require a similar longterm effort.

Protecting against climate change threats to US national security will require broad-based action well before a conflict. In order of increasing effectiveness, the spectrum of national climate action responses include kinetic military operations, aid after the effects of climate change are manifest, adaptation (encouraging permanent changes in the affected area to cope with climate change effects), and to develop indigenous resilience in anticipation of climate change impact. To the extent that the nature, impact, and location of destructive influences of climate change can be reasonably predicted, the most effective, least costly actions to protect US national security lie at the development of indigenous resilience end of the response spectrum; the least effective, most costly lies at the kinetic military end. Setting the right priorities, and effectively engaging the right government entities, will require broader and deeper attention. Still, there is increased sensitivity to the potential impacts of climate change and the need to address them.

Former Secretary of Defense Robert Gates addressed the importance of recognizing the Department of State and USAID as elements of national security. Secretary of State Hillary Clinton spoke of the importance of migrating development functions back to the Department of State and USAID. The 2010 US National Security Strategy assigns the responsibility for addressing the climate change threat to national security to both DOS and USAID. No senior-level Pentagon official has been assigned responsibility for the DOD interest in climate change adaptation. While there is a general recognition of the importance of the impacts of climate change to national and international security, there is a need for clear top-down guidance to proactively address climate change adaptation before it leads to conflict or instability. There is a need for clear assignment of an individual and organization as the coordinating agency in the US government responsible for climate change adaptation at the regional level.

National leadership, interagency coordination, and the strategic documents that govern policy development in Washington need to be further strengthened. Still, at the regional level, the framework for cooperation on climate change capacity building exists. The regional missions of USAID and the Department of State Regional Environmental Hubs are currently focusing on sustainable development and environmental diplomacy in ways that address such climate change related issues as water resource management and adaptive agricultural practices. These programs have successfully sought to develop host nation and regional climate change resilience and should be given consistent combatant command support. Providing the support of the combatant command engagement activities would multiply the effectiveness of these programs and could markedly change the conditions for populations impacted by climate change. Several nations are currently facing new and growing environmental challenges, many of which require regional cooperation to solve. Bringing nations together in a region to work on a common environmental problem—a common threat—can advance US interests in ways that go far beyond the scope of the environmental issue itself. The government's commitment to a regional strategy complements our bilateral and multilateral diplomatic environmental efforts.

During the Clinton Administration, the Department of State created the Regional Environmental Hub program as the flagship for its environmental diplomacy efforts. It was created with the belief that, "Bringing nations together in a region to work on a common environmental problem—a common threat—can advance *US* interests in ways that go far beyond the scope of the environmental issue itself."[134] These offices have thrived and, during subsequent administrations, have consistently brought multilateral cooperation to environmental issues that threaten to destabilize regions. In 2009-2010, the Department of State Regional Environmental Hub for Central America and the Caribbean partnered with the US Southern Command to conduct two climate change and security events to determine climate change adaptation threats to the region and to develop responses. The second conference was the region-wide, Central America and the Caribbean Climate Change and Security Conference: from Strategy to Action. There are three Regional Environmental Hubs in Africa and two in South and Central America.

In the 1990s the Department of Defense established the preventive defense concept to avoid costly conflict and successfully sought interagency cooperation in its efforts to "mitigate the impacts of adverse environmental actions leading to international instability."[135] Since then, the regional combatant commanders have used environmental security as an engagement vehicle and have worked closely with the Hubs to build cooperative relationships among regional states and military support for civil authority and democracy. DOD cooperation with partner countries has been regularly supported by agencies such as the US Agency for International Development, US Geological Society, the Environmental Protection Agency, and the Department of the Interior. These build partner capacity and capabilities to address environmental security issues and promote stability. It is important to understand that this international interagency cooperation is ongoing and already addressing the security dimensions of many climate change issues.

Findings on Strategic Concepts for Addressing the Challenges

The least expensive way to deal with threats to US national security will be through anticipatory and preventative actions using primarily indigenous resources.

The United States needs better insight into the political, economical, and geographical impacts of climate change.

US national security threats will require government-wide, coordinated outreach in training and education, planning, engineering, agriculture, health, justice, and military areas.

No single person or organization in the US government has been assigned coordinating agency responsibility for climate change adaptation or coordinating interagency efforts to address issues at the regional level.

There is a need for a strategic level, interagency process to identify climate change adaptation hotspots, determine what resources each agency can provide, and synchronize cooperation to create climate change resilience.

Climate change effects, particularly those related to water and food security, can erode the legitimacy of fragile states and create the conditions terrorists and extremists seek to exploit. Therefore, they are significant factors in combating terrorism.

The United States can best use its informational and technical capabilities to build and support local resources to anticipate and adapt to climate change impact.

The United States has demonstrated capabilities to deal with issues similar in scope and similar to the potential impacts of climate change, e.g., the Agricultural Extension Program that led a farm revolution and transformed the food supply, the Public Health Service that fostered changes in sanitation and food and water processing that virtually eradicated common childhood epidemic diseases, and the US Army Corps of Engineers and the Bureau of Reclamation, since the 1890s, that altered the US riverine transportation system and enabled growth in the water- starved West.

The United States has demonstrated a capacity for training in the foreign military training programs that could be expanded to provide education and training most relevant to adapting to climate change, e.g., hydrological, civil engineering, construction, agricultural, biological, and medical training.

Disaster Preparedness and Climate Change

As Hurricane Katrina and the earthquake that led to the fall of the Somoza government in Nicaragua made clear, natural disasters that expose a government's lack of preparedness have quantifiable effects on governmental legitimacy. Climate change will have a disruptive effect on state systems, putting at risk the resource base and sustainability of the government. Reducing the risk and responding to resulting natural disasters are emerging challenges for local, regional, and state governments. Lack of preparedness for environmental challenges, such as those in Tunisia, Egypt, and flood-ravaged Pakistan can multiply tensions from existing grievances and lead to instability. Resilience, sustainability, and preparedness are essential to avoiding political instability and important to any national security community effort to mitigate the regional effects of climate change.

It is useful to conceptualize the role of the national security community in addressing this destabilizing issue as creating climate change resilient communities. NOAA was tasked by Congress in 1994 to assess tsunami awareness and preparedness for parts of the United States. As a result of their analysis and research, NOAA developed a concept for mitigating the damage of tsunamis. Called Tsunami Resilient Communities, it was created "to provide direction and coordination for tsunami mitigation activities in the absence of a disaster."[136] Recognizing that no effort would be successful without the support of local communities, NOAA designed a plan to leverage planning, education, and awareness to minimize losses and reduce fatalities and property damage. The seven variables of resilient communities are designed to enhance national, state, and local capabilities by determining the threat, preparedness, timely and effective warnings, preparation public outreach and communication, research, and international coordination. This concept could easily be adapted to climate change adaptation and would provide a clear end state for national security community efforts.

The concept of resiliency is in use by the United States Pacific Command's Center of Excellence for Disaster Management and Humanitarian Assistance (COE) as a framework to guide its efforts in dealing with the near-term effects of climate change. COE, now in its second decade, believes that the concept of societal resilience provides the best form of disaster preparedness to avoid unnecessary disaster response missions. It works with the interagency community, combatant commands, the international community, and NGOs on disaster preparedness and response efforts and drew upon that experience to draft its climate change resiliency concept. As presented by LtGen (Ret) John F. Goodman, COE Director, during his address to the recent Pacific Command Environmental Security Conference, the pillars of the climate change disaster resiliency concept are:

- **Societal capacity.** In order to confront the challenges posed by disasters, the community must have a developed public health system, diverse infrastructure, humanitarian assistance capabilities, and robust disaster preparation.
- **Knowledge assets.** An equitable education system, hardened communication and information exchange mechanisms, and inviting environment for research and innovation are critical to developing the intellectual capital to ensure resilience.
- **Resource independence.** A robust economy and established sustainable development and resource management practices enable a community to withstand and recover from the financial and environmental disaster threats.
- **Community cohesion.** A community that shares common values and objectives while respecting diversity and striving for social parity can better work collectively to resist the effects of disasters and rebuild in their aftermath.
- **Good governance.** A resilient community requires accountable leadership, just legal and regulatory codes, appropriate security and social protection mechanisms, and the ability to assess its risks.

COE's climate change adaptation program, founded on resiliency and its supporting framework, reinforces many of the objectives of USAID's sustainable development concept and the intent of the National Security Strategy. The success of a regional security organization such as COE in developing such a concept in cooperation with other interagency and international organizations, and applying this concept in its educational and operational activities, demonstrates the feasibility of a broader government approach to climate change adaptation. This type of coordination normally occurs after the onset of the crisis, as was the case in Iraq and Afghanistan. The provincial reconstruction teams and the agribusiness development teams have successfully worked to reduce the conditions of drought, water, and dry land agriculture that extremists seek to exploit in Afghanistan. These conditions mirror conditions already causing tensions in other regions from climate change effects. At the operational and tactical levels, much the same as with embassy country teams, interagency cooperation is possible and essential to success.

At the strategic level, success is also possible. Priorities in the national security community are recognized with the creation of an Interagency Policy Committee that develops policy options, assigns responsibility, and coordinates actions. Climate change adaptation is an important national security issue that affects regional stability and US national security interests, warranting this level of priority. With top level guidance and priority, agency strategic documents, such as the National Military Strategy or the Department of State and USAID Strategic Plan could speak to the importance of interagency cooperation and create regular cooperation and synergy among valuable programs that are already in place.

Without this top level coordination and priority, strategic documents of the national security community may fail to stress the importance of climate change adaptation in a meaningful way, and the operational and tactical levels of these organizations will only undertake climate change adaptation if lower-level leadership deems it important.

Moreover, the opportunity to promote multilateral cooperation in addressing a growing threat to regional stability and to undertake confidence-building measures that could lead to peace building could be lost.

Findings on Disaster Preparedness and Climate Change

Governments unable to meet challenges placed on their political systems by the effects of climate change are at greater risk of failure; this is particularly important in those areas where fragile states are pivotal to US national security interests.

Proactively reducing risk and responding to the increased natural disaster potential from climate change can best be achieved by promoting the concept of climate change resilience as a common theme to guide regional interagency activities.

Whole of Government

The government has numerous organizations which can contribute to both the understanding of climate change and actions to address near- term response and adaptation.

US Government Organizations

The Department of State has established a Special Envoy for Climate Change, and ambassadors lead our efforts abroad through the country team.

USAID has the Famine Early Warning System which provides indicators to complement the Intelligence Community's monitoring capabilities. The Senator Paul Simon Water for the Poor Act directs improving access to water supplies and sanitation, hygiene, and water management.

The Central Intelligence Agency has a Center for Climate Change and has established the Measurements of Earth Data for Environmental Analysis program to inform the scientific community of classified data.

The Department of Agriculture has a Foreign Agricultural Service which helps educate farmers worldwide how to improve productivity.

The Department of Energy manages an extensive laboratory system employing scientists of the highest caliber and an office of Energy Efficiency and Renewable Energy which contributes lessons.

The Environmental Protection Agency has a Climate Program Office and establishes policies and standards to mitigate climate change impacts.

The National Oceanic and Atmospheric Agency collects and publishes critical data on sea level and atmospheric changes.

The Jet Propulsion Laboratory can provide data on: 1. Carbon-agriculture-forestry and land management; 2. Precipitation and fresh water; 3. Sea level rise and coastal surveillance; and, 4. Geographic hazards-earthquakes, volcanoes, and tsunamis.

US Geological Survey uses a whole systems approach to understand global change. It documents and models past and present climates and environmental change through geological, biological, and hydrological processes, and has regularly supported the combatant commands.

The National Academy of Sciences and the National Research Council publish critical studies by eminent scientists on the causes and effects of climate change.[137]

The President's Office of Science and Technology Policy through the President's Committee on Science and Technology provides advice and guidance on the science of climate change.

Other US Organizations

In addition to governmental organizations, there are numerous organizations which can provide assistance and advice. They include The Nature Conservancy and the American Red Cross, as well as dozens of local and university groups which provide research and adaptation projects. NGOs provide a wealth of expertise and assistance from both with the United States and internationally.

International Organizations and Allies

Outside the United States, there are numerous efforts underway to address climate change and its impacts. The United Nations provides both military and humanitarian response to mitigate the results of aggression and disasters. The United Nations Environmental Program monitors and advises on the impacts to our environment. The World Bank and the African Development Bank monitor and provide financial assistance to adapt to the impacts of climate change. Britain recently cosponsored a SOUTHCOM regional Climate Change and Security Roundtable in Colombia aimed at building military capacity to address the effects of climate change.

Findings on Whole of Government

There are US government organizations and international organizations available to:

- Provide indicators of pending disasters
- Support efforts to proactively build adaptation and resilience capacity
- Assist in the response to the effects of climate change
- Provide scientific and engineering support to understand and help mitigate change

There is no central organization to assist agencies in understanding what resources are available or to coordinate their efforts.

Recommendations on Roles of the National Security Community

The Director of National Intelligence should:

- Establish, within an appropriate agency of the Intelligence Community, an intelligence group to concentrate on the effects of climate change on political and economic developments and their implications for US national security
  - An important focus of this effort should be to project human security changes that could develop into national security issues.
  - This group should make extensive use of open sources, seek to cooperate with other domestic and international intelligence efforts, and report most of its products broadly within government and nongovernment communities.
- The intelligence group should commission the Central Intelligence Agency's (CIA) Center for Climate Change and Security to produce

an assessment of regional climate change hotspots that threaten human security and governmental legitimacy and exacerbate existing tensions. They should use this assessment as a confidence-building measure to promote communication between antagonistic peoples or states. This document should be the basis for interagency cooperation at the strategic and regional levels.

The President's National Security Advisor, in conjunction with the Council on Environmental Quality, should establish an interagency working group to develop:

- Coordinated climate change policies and actions across US government entities
- A whole of government approach on regional climate change adaptation with a focus on promoting climate change resilience and maintaining regional stability
- The President's National Security Advisor should continue to emphasize strategic interagency documents, such as the DOD Strategic Guidance which details the link between climate change effects and the underlying conditions that terrorists seek to exploit and should direct relevant organizations to consider this relationship in developing their regional plans.

The Deputy Secretary of State and the Deputy Secretary of Defense should:

- Follow the example of the successful foreign military training assistance program to fashion education and training programs in the fields most relevant to adapting to climate change, e.g., hydrology, civil engineering, construction, agriculture, biology, and public health.
- Make conflict avoidance a priority of foreign assistance (including security assistance and foreign military sales), development, and defense concept development, and planning.
- Develop a strategic communication message that links water and food security and increased storm intensity to regional stability and US national security.

5. Role of the Department of Defense

Guidance Shaping Department of Defense Efforts and Activities

The National Security Strategy describes dangers arising from climate change: "...new conflicts over refugees and resources; new suffering from drought and famine; catastrophic natural disasters; and, the degradation of land across the globe."[138] It further emphasizes that efforts to mitigate and minimize the impact of these changes need to be pursued both at home and abroad through international cooperation. There will be significant impacts on DOD.

President Obama, in accepting the 2009 Nobel Peace in Oslo stated that, "...it is not merely scientists and activists who call for swift and forceful action—it is military leaders in my country and others who understand that our common security hangs in the balance."[139] This point was included in the 2010 Quadrennial Defense Review, the first DOD strategic guidance document to give thorough treatment to the issues of climate change and energy. Thus, for the first time, the two documents that set the framework of defense policy guidance explicitly called for the need of DOD to address domestically and internationally a full range of issues associated with climate change. This guidance has since been incorporated into other DOD documents.

DOD clearly has significant roles in dealing with climate change. For its internal needs, DOD will need to assess how climate change can impact readiness by affecting or altering:

- Existing and planned military facilities and equipment both at home and abroad
- Training, exercises, and deployment of these forces
- The health and safety of military personnel
- The frequency, location, and types of military operations, the need for new or expanded training, and new equipment needs

Externally, DOD must prepare to support the climate change initiatives of the US government as a whole. DOD can, and will, be expected to provide critical support to interagency climate change efforts to direct near-term activities and adaptation towards conflict prevention. DOD will also need to play a lead role in military to military cooperation with other nations to enhance their capacity and resilience to deal with the impacts of climate change. Combatant commands will need to integrate climate change near-term response and adaptation into their theater security cooperation programs and campaign plans. Tangible steps may include identifying regional climate change threats to stability, building the capacity of regional militaries to support civilian authority in addressing these threats, assisting in monitoring and data collection, and engaging with foreign militaries on disaster preparedness.

DOD will need organization to address the full range of international climate change-related issues and their impact on the evolution of DOD's missions. Currently multiple DOD offices are addressing climate change and energy issues with a wide range of perspectives and with limited unifying guidance. This fragmented approach is inadequate to the need. The need is for clarity in responsibility and accountability. The need is for a lead office in the Office of the Secretary of Defense (OSD)—an office designated as the centralized DOD point of contact to serve as the coordinating authority within DOD and to act as the designated DOD representative or interface in relevant interagency activities. Similarly, each of the Services and the Joint Staff need to designate specific leads within their organizations. The current general guidance needs to be translated into specific requirements to provide executable orders to the Services and the combatant commands to address these issues more systematically and comprehensively.

The National Security Strategy, the Quadrennial Defense Review, and other documents have stressed that climate change impacts national security. It is important for all elements of the DOD to understand that climate change can, and will, impact its ability to carry out DOD missions now and in the future. The issues impact the combatant commands, the military departments, and at least some elements of the Joint Staff: J2 for providing required intelligence information, J4 for logistics and installations, J5 for assessing regional stability and country-specific impacts, and J8 for providing the required resources. There are ongoing Services' initiatives relevant to climate change and disaster risk reduction and the impact on their security interests and operations.

The US Navy has taken a forward-leaning approach, initially focusing on increased Arctic operations resulting from climate change. Navy attention has since expanded to include assessments of the potential offered by increased partnerships with other armed forces, the impacts of sea level rise on installations, and the potential for increased humanitarian assistance/disaster relief missions. The Oceanographer of the Navy heads the Task Force on Climate Change and issued a Climate Change Roadmap signed by the Vice Chief of Naval Operations in May 2010. This roadmap outlines the Navy's approach to observing, predicting, and adapting to climate change in the 2010-2014 time frame. It includes incorporating climate change impacts on national security into war college courses and strategy guidance documents; beginning to define the requirements of a next generation operational and climatic environmental prediction capability; including climate change considerations in training and planning; and pursuing international cooperation to enhance the Navy's capacity to assess, predict, and adapt to climate change.

The Marine Corps' expeditionary energy and water program focuses on expeditionary solutions for sustainable energy, which could have applicability to small or developing nations addressing climate change- related impacts, particularly with respect to those with limited infrastructure.

The US Army's climate change-related initiatives have resided principally with the USACE. On the domestic side, the USACE has been tasked to examine the effects of sea level rise on its installations in the continental United States. Internationally, the USACE has been working with other nations on water-related issues, such as water availability, conflict resolution scenarios, and water resource operations infrastructure development in foreign operations.

Air Force capabilities include its significant suite of meteorological data collection, atmospheric assessments, observational capabilities, and satellites. In addition, the Air Force provides air transportation and civil engineering units vital to disaster response missions.

The National Guard offers other important assets that can be drawn upon by the combatant commands to support activities in their AORs. Notably, the State Partnership Program has established relationships between various state National Guards and 62 countries throughout the world. Drawing on the expertise that its personnel bring from their civilian occupations, the National Guard supports US national security goals and assists in the achievement of respective theater security cooperation and individual country campaign plan objectives. It is able, for example, to help build capacity in partner nations in areas such as disaster preparedness, resource management, and other topics related to climate change effects.

A number of existing programs and activities support the Services' and combatant commands' needs for climate change-related initiatives. As noted above, the Navy Task Force on Climate Change and USACE have important vehicles for undertaking such work. In addition, both the Office of the Secretary of Defense for Policy (OUSD(P)) and the geographic combatant commands can draw upon the Defense Environmental International Cooperation Program, managed by the Environmental Readiness and Safety office in Office of the Deputy Under Secretary of Defense for Installations and Environment. DEIC is designed to support engagement activities with other nations' defense institutions.

As an example of work at the combatant command level, the AFRICOM Environmental Security Program, working with US government and international partners, is already in the process of implementing a series of climate change-related activities, such as coastal erosion and water security assessments, incorporating the participation of civilian and military stakeholders alike. The Africa Partnership Station represents another valuable program to help build maritime security capacity in a host of African nations. The themes addressed under the program can easily be tailored to include those related to climate change and disaster risk reduction.

Findings on Guidance

DOD has unique capabilities and resources to help deal with climate change and disaster risks to include specialized expertise (e.g., engineering, hydrology, logistics, air and sea lift, and innovative research).

DOD's well-established, long-term planning capability and the engagement programs of the combatant commands can contribute to providing broader and higher quality climate change information to the domestic and international community.

DOD capacity needs to be presented with sensitivity to other agencies with more limited resources but with authorities and mandate to lead the effort.

DOD currently engages in a range of approaches to engaging with the interagency on climate change. Individual offices bring individual perspectives and equities to their respective interagency discussions but have not systematically recognized climate change adaptation as a significant regional requirement for stability.

The Office of the Under Secretary of Defense for Acquisition, Technology and Logistics (OUSD(AT&L)) represents DOD in interactions with the Council on Environmental Quality.

The Deputy Assistant Secretary of Defense for Partnership, Strategy and Stability Operations interacts with National Security Staff on matters pertaining to development.

The Joint Staff J5 Directorate, Strategic Plans and Policy, and the Deputy Assistant Secretary of Defense for Strategy work with National Security Staff on matters of strategic planning.

The appreciation across the elements of DOD that climate change and disaster risk reduction have important implications for its roles and missions varies across Services and across the combatant commands.

There are parallels between climate change today and environmental security in the 1980-2000 time period. OSD developed an understanding of, and competency in, environmental security which was coordinated with the relevant regional offices within OUSD(P) and was available to the combatant commands and OUSD(P) in their bilateral and multilateral engagement activities.

Climate change is currently having a major impact on the demands of military operations in the Arctic. DOD will need additional capabilities to meet the demands of the expanded Arctic mission. For example, climate change is creating an unfunded mandate for additional ice breaking capability, but ice breakers are under the authority of the National Science Foundation, not under military authority.

Recommendations on Guidance

The Deputy Secretary of Defense should:

- Establish a DOD-wide coordinating policy board for climate change impacts on national security. This board's functions should include:
  - A coordinating role on climate change information from the strategic and operational perspective. This would include assessing implications for the force structure, deployment options, etc.
  - Compiling and assessing climate change effects information across the geographic combatant commands to identify implications for regional stability and the development of global and regional foreign military assistance programs.
  - DOD's interagency representative for climate change adaptation matters.
  - Serving as the focal point for information, web-enabled, that can be accessed by other Office of the Secretary of Defense (OSD) offices as well as the Joint Staff, Services, and combatant commands.
- Expand the authorities of the Operational Energy Plans and Programs Office to include operational climate change issues.
- Direct the establishment of a program of climate change adaptation pilot projects in concert with related programs at USAID and other

agencies to identify, solicit, and fund pilot projects focused on specific adaptation sectors and locales (e.g., management of regions or villages in Africa and Central Asia). Examples of pilot projects and suggested activities might include, but are not limited, to:

- Embrace and augment the World Climate Research Program CORDEX for one of the sub-regions in Africa. Apply CORDEX in concert with an assessment activity similar to the European PRUDENCE project.
- Extend the observational, modeling, and synthesis assessment capabilities applied today in the United States in the Upper Colorado River Basin to a priority water resource district in Africa, perhaps linked with the Nile Basin initiative.
- Apply coastal hot spot pilot projects focused on offering local-scale risk assessment and planning for integrated sea level and storm impacts on the coupled water-energy-waste resources and physical infrastructures for megacities such as Lagos, Karachi, and Dahka.
- Engage the USGCRP, international research programs, DOD commands and their in-country security partners, and international aid agencies such as USAID in identifying opportunities to share climate change-related information and bringing more visibility into stakeholders' activities.
- Focus on near-term, achievable, and measurable goals to develop and demonstrate end-to-end threads of core information systems while incrementally building in-country capacity and competence.

OSD, Office of the Under Secretary of Defense for Policy and the Director, Joint Staff should direct development of a DOD strategic roadmap for climate change-related efforts that builds on the framework laid out in the US Navy Climate Change Roadmap to:

- Ensure that the guidance to the combatant commanders, once signed, is considered to be adequate by the Services and combatant commands for translating the broad-level guidance offered in the Quadrennial Defense Review into actionable requirements.
- Direct that combatant command missions include non-combat support to address serious climate change-induced US national security vulnerabilities.

The Deputy Under Secretary of Defense for Installations and Environment should assemble an inventory of critical facilities and infrastructure to include an assessment of vulnerability to climate change effects and the means to adapt.

The Director, Joint Staff should:

- Create a holistic approach to climate change, integrating efforts of its relevant directorates: J2 (Intelligence), J4 (Logistics), J5 (Strategic Plans and Policy), and J8 (Force Structure, Resources, and Assessment Directorate).
- Require that climate change and disaster risk reduction be integrated into training and exercises.

The Secretaries, Chiefs of the Services, and heads of defense agencies should:

- Better integrate climate change and disaster risk reduction considerations into exercises, training, and educational materials.
- Establish metrics focused on risk reduction to minimize the impact of climate change on military and support operations, forces, programs, and facilities.
- Develop guidance to ensure climate change resilience in DOD project designs and construction by incorporating climate change risk into design standards for facilities and installations, with emphasis on the elements related to energy intensive and water intensive uses.

The Secretaries and Chiefs of the Services should:

- Assess the Services' engineering organizations and the cost- benefits of using them in assisting climate change adaptation.
- Utilize military to military engagement opportunities with coalition partners to enhance resilience to climate change impacts and disaster risk reduction capacities. In so doing, they should expand consideration of roles for the National Guard and reserves. (For example, knowledge of traditionally non-military skills needed to respond to climate change threats is often found in the reserves.)
- Examine tasking authorities for domestic and international response to natural disaster or other disaster risk response situations. For example,

the National Guard could bring important assets to an international disaster, as it already does in responding to domestic disasters.

United States Northern Command, with support from the Navy and Coast Guard, should identify the assets that will be needed to operate in the Arctic to include communication assets, personnel training, ice breakers, and other equipment.

Combatant Command Roles, Responsibilities, and Capacity

Each combatant command has unique missions and priorities. However, there is a universal need for each combatant command to consider the potential climate change and disaster risk reduction impacts on their readiness to meet mission needs in their AOR. One example of combatant command information requirements, AFRICOM needs detailed assessments within its AOR of climate and natural resource vulnerabilities, such as weather, food, and water. This information is crucial for defining AFRICOM's operational interests and requirements and also for identifying priorities in its interactions with African nations. In effect, such assessments are analogous to doing a national impact assessment at the regional level. These assessments would:

- Identify areas, countries, and regions where the impact of climate change on stability and US security interests is greatest.
- Identify the nature and potential scope of the most likely impacts.
- Identify what capacity building measures should be undertaken to address these impacts.

This information is essential for AFRICOM to develop its priorities for resource allocation. The criteria for funding would be a combination of the country's strategic importance to US security interests, the country's own capacity for addressing the impacts, the potential severity of the impacts, and possible spill-over effects.

- Each combatant command should include in its objectives:
  - Developing and enhancing host nation military capacity to build resilience to climate change effects
  - Raising the level of awareness of these issues with other partner nations and their militaries

- Taking into consideration capacity infrastructure interests of the United States

Again taking AFRICOM as an example, it has used vehicles such as the Africa Partnership Station, African Endeavor, the DEIC program, the National Guard State Partnership Program, and the Humanitarian and Civic Assistance program to:

- Enhance capacity on land, at sea, and in the air; promote interoperability
- Provide medical and veterinary assistance; prepare for crisis response
- Plan and prepare for disaster relief

This is not an exhaustive list of the programs AFRICOM has utilized or of its engagement activities; rather, is it offered as an illustration of existing resources that can support AFRICOM efforts to prevent the destabilizing effects of climate change.

United States Special Operations Command has important engagement activities that can similarly incorporate climate change and disaster risk reduction considerations. This is particularly important given the implications of climate change effects for fragile state legitimacy and combating terrorism. United States Transportation Command will need to prepare to respond to disaster relief missions and also to supply forces in the Arctic as that mission evolves.

The ultimate objectives for each combatant command are to build the capacity to operate successfully in a climate-changed milieu and the capacity of host nation militaries to address climate change effects that threaten regional stability.

Findings on Combatant Command Roles, Responsibilities, and Capacities

Climate change and natural resource issues should be viewed as cross-cutting issues rather than predominantly the concern of the J4 (Logistics). Climate change will have impacts across the command to include; planning, logistics, training, and relationship-building with nations in the AOR.

An interagency approach is essential to addressing climate-change related topics as many of the themes are outside the mandate of DOD-specific responsibilities.

SOUTHCOM and AFRICOM offer useful models for addressing climate change-related issues. United States Pacific Command has opted to use its Partnership Office (J9) to lead on these issues. Whatever the approach, these issues cut across the J-staff structure.

The combatant commands need focal points in OSD and the Joint Staff.

Regional US military commands have managerial and technical expertise and access to resources, like transportation, necessary to support and, when appropriate, lead US and international efforts to ameliorate climate change threats to US national security.

There are resources outside those normally tapped by the combatant commands to help address climate change and disaster risk reduction themes, e.g., DOD's Minerva Initiative, engineering services, the National Guard, and OSD-run programs. For example, much of the US military civil affairs expertise resides in National Guard and reserve units.

Recommendations on Combatant Command Roles, Responsibilities, and Capacities

The geographic combatant commands should:

- Identify early warning indicators for those areas critical to DOD's mission set.
- Incorporate the guidance from the Quadrennial Defense Review and classified guidance to the combatant commanders on energy, security, and climate change into theater campaign plans.
- Create a demand signal by articulating the need to understand the implications of climate change and resource scarcities in their region to support their campaign plans.
- Include in their theater campaign plans energy, food, water, and disaster risk reduction strategies and plans for reducing vulnerabilities within their respective AORs.
- Harness more systematically resources beyond the traditional combatant command structure, to include the National Guard and its State Partnership Program, service engineering units such as USACE and Naval Facilities Command, and OSD-led programs such as DEIC and the Strategic Environmental Research and Development Program.
- Conduct systematic regional or even more localized impact assessments to identify trends and where their resources should be focused. To this end, each should request that the CIA's Climate

Change and Security Center provide a report on climate change effects and hot spots in their respective areas of responsibility.

- Programs such as the Strauss Center's program on Climate Change and African Political Stability, funded through DOD's Minerva Initiative, could also be utilized in such undertakings.
- Include as a Tier 1 objective enhancing the capacity of host nation militaries and civil response readiness groups to plan for, and respond to, natural disasters (e.g., floods, coastal storm surges, and droughts).
- Integrate into their humanitarian assistance/disaster relief and other exercise plans additional climate change-related aspects. These exercises should include interagency activity.
- Promote the concept of coordinated management of shared natural resources like water.

APPENDIX A. CLIMATE INFORMATION SYSTEM NEEDS

As introduced in Chapter 1, the current collection of observational and model assets while important for conducting exploratory climate science do not constitute a robust, sustained, or comprehensive resource for generating actionable climate forecasts. Developing adequate climate projections will require contributions from, and cooperation within, the US government of NOAA, NASA, US Geological Survey, and the Central Intelligence Agency; the Departments of Agriculture, Defense, Energy, and State; and others with climate, geographic, economic, social, and political skills. Since much of this expertise lies outside the government, universities, the private sector, and NGOs will also need to be involved. Understanding a climate-changed future will require the richest possible effort, encompassing the world's best expertise. As with any mystery[140] the most effective way to tackle understanding it may be to treat it, for the most part, as an open question, transparent to all engaged in its study. Compartmentalizing climate change impact research can only hinder progress.

Climate Change Risk Management

Effective planning and execution of climate change responses are primarily exercises in risk management. This is a process (Figure A-1) involving the collection and evaluation of information, quantification of uncertainties, evaluation of risks (likelihood and impact), and response options, decision-making, and iterative re-evaluation.[141] Climate change risk management shares many of the information resources and decision- making processes involved in other decision-making processes already employed by governments, businesses, and other organizations.[142] It shares their limitations and also introduces additional challenges with implications on the supporting information resources:

- Increased climate and ecosystem variability and extremes
- Climate/ecosystem threshold crossing events or tipping points
- Coupling across multiple scales (teleconnections)
- Interaction between human response efforts

In the absence of climate change, traditional risk management often relies on relatively steady-steady assumptions with regards to environmental variability and extremes. Risk management for droughts, floods, hurricanes, and geohazards in most cases relies on information based on current observations and relatively recent historical trends from decades to centuries. The difficulty of managing risk for relatively mundane hazards was vividly demonstrated in the March 2011 events in Japan following a magnitude 9.0 earthquake, tsunami, and subsequent cascading failures of several nuclear reactors.[143] The potential of climate change to modify the steady background state and/or to significantly increase the frequency and intensity of climatic extremes such as heat waves, drought, heavy precipitation, storm surges, and tropical cyclones will likely stress the ability of existing information resources to accurately represent the distribution of those events. Climate change also introduces the possibility of abrupt, potentially irreversible threshold-crossing events or tipping points, in which the Earth's climate system undergoes a non-linear response, essentially shifting into a qualitatively different and poorly characterized new stability regime,[144] a topic explored further in Appendix B. Special Topics.

Climate change involves direct coupling across a range of spatial and temporal scales that both overlaps and exceeds the boundaries of traditional risk management processes. Conventional risk assessment for tropical cyclones (hurricanes) and other severe weather currently leverages observations and models with global scope to produce local/regional scale forecasts and nowcasts. However, weather forecasts and supporting observations focus on timescales ranging from hours to days whereas climate forecastor projection time horizons span seasons to decades. Recent disasters such as the EyjafjallajOkull (2010) volcanic eruption, the Sumatra (2004) tsunamis, and Japan (2011) earthquakes offer stark reminders of the far-reaching impacts of localized and relatively short-lived events on global air transportation and coastal populations. However, such events are driven by geologic rather than climatic processes. Thus, risk management for them, to some extent, will remain invariant to future climate change.[145]

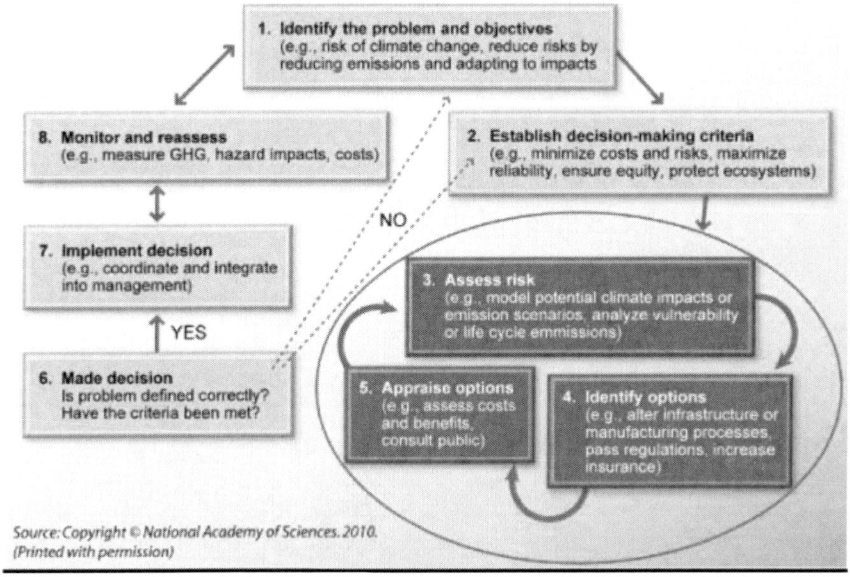

Source: Copyright © National Academy of Sciences. 2010. (Printed with permission)

Figure A-1. The concept of an iterative, adaptive risk management process for climate change.

The El Niño Southern Oscillation is a better example of a climate teleconnection, a strong, low-frequency (e.g., ~months) correlation with planetary scales. There are numerous examples in the literature indicating the presence and importance of such large-scale coupling mediated via ocean-atmosphere circulation and leading to remote-region impacts. These include tropical Pacific-North American patterns associated with ENSO, lower-frequency variations between the Pacific Ocean/ENSO, the Asian monsoon and/or Sahel rainfall, and the influence of the Indian Ocean on European/Atlantic climate.[146] Other teleconnections span the boundaries of geophysics and economics, such as CO_2 emissions embodied in international trade,[147] the lateral transport problem faced by carbon accounting systems with direct implications for the atmosphere and climate forcing.

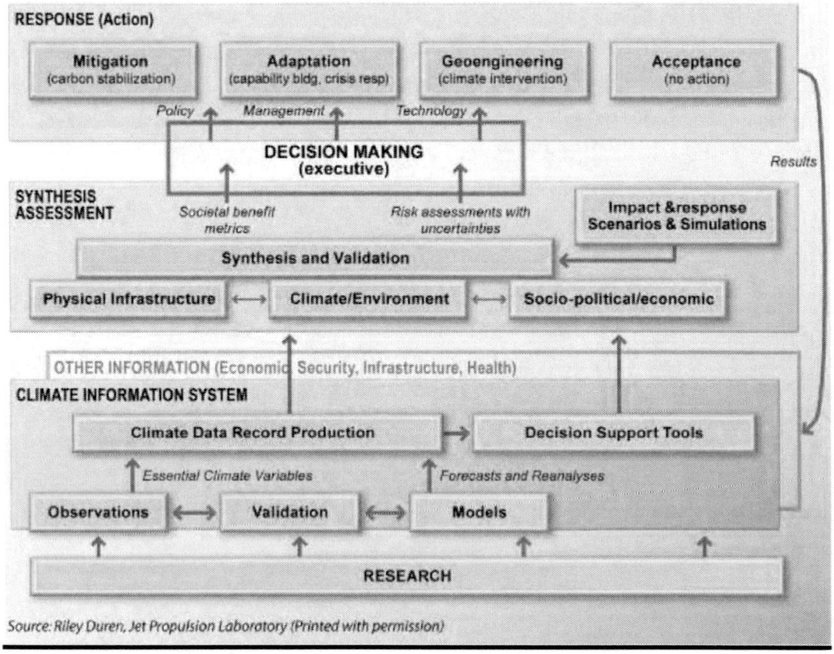

Figure A-2. Notional framework for implementing climate change risk management.

Finally, the response space for climate change risk management is large, spanning mitigation (efforts to minimize climate change by stabilizing greenhouse gas emissions), adaptation (proactive and reactive efforts to minimize the societal impacts of climate change), geoengineering (a wildcard option described further in Appendix B), and acceptance (no action). Furthermore, the interactions between these response options are complex and uncertain, both in terms of decision-making (economic and policy considerations) and the effects on the Earth system.[148] For example, in the pan-tropics, public and private landowners are already forced to choose between preserving a stand of forest to receive a carbon offset credit under the United Nations program to Reduce Emissions from Deforestation and Forest Degradation versus planting biofuels versus planting grain for food, with implications on the carbon-cycle (contributing to climate change), ecosystems, watersheds, and agriculture.

In practice, given the wide range of decision makers and complexity of the supporting information, there will be no one-size-fits-all climate change risk management system. A notional framework for climate change risk management in Figure A-2 is introduced to offer some context from a US government perspective for the following evaluation of information systems and related processes. The framework is grounded in various information systems which offer input to assessment processes.

The climate information system would use a combination of sustained observations and models to monitor, forecast, and reanalyze data associated with essential climate variables. A rigorous validation function will be essential to quantifying uncertainties and managing errors. Climate data records, including trends, will be produced, archived, and disseminated. Decision support tools present climate data in formats amenable to interpretation by non-climate scientists and risk assessors. The foundation for, but to some degree separate from, the operational climate information system is a research element that remains flexible to discovery and ensures that the latest understanding of climate processes is infused into operational information. The research program will contribute to, and benefit from, the outputs of the climate information system.

Synthesis assessment is a collaborative multidisciplinary process by which subject matter experts in the natural sciences, social sciences, economics, and policy translate the outputs from climate and other information systems into societal benefit metrics and actionable information for decision makers. Synthesis assessment includes the generation and simulation of scenarios to help evaluate risks spanning the option-space of response actions which are mitigation, adaptation, geoengineering or non-action which is acceptance. In some other frameworks, synthesis assessment is treated as a component of a climate information system but is presented separately here to emphasize the need to infuse non-climate information into the assessment process.

The decision-making executive function integrates the knowledge provided by the information systems and assessment process and takes action in the form of policy, management, deployment of technology, and other resources. Ideally, this executive function would serve as the overarching director and integrator of the climate change risk management framework. In a well-constructed risk management framework, the information systems and assessment function would receive requirements from the decision-making executive, starting with well-posed questions about the range of possible responses flowing down to what information is needed to support risk assessment.

Some potential contributing factors and consequences are described in the following sections.

Findings on Risk Management

Effective climate information systems will need to identify changes in the distribution of extreme events driven by changing climate, including updated likelihoods, impacts, and uncertainties and provide for frequent reassessment.

Risk management frameworks that have traditionally relied on relatively static or weak global connections will be challenged in the future to quantify climate driven changes in the modes of variability and strength of geophysical and socioeconomic teleconnections spanning the Earth.

Information systems and assessment processes need to enumerate and quantify the potential interactions across and between response options towards understanding the risks and benefits of taking a given action.

While there are related efforts underway by various organizations such as NOAA and the US Global Change Research Program (USGCRP), no systematic effort to define requirements, architecture, and implementation plan for a climate risk management framework and supporting information systems and assessment process has been completed by the US government or other entities.

In the absence of a focused effort to understand key needs, an infusion of new funding to develop a truly operational climate information system, and a mandate to address regions outside the United States, the tentative NOAA Climate Service and related programs may prove insufficient to meet risk management needs.

Climate Information Systems

Attributes of an operational climate information system include, but are not limited, to:

- Reliable, sustained climate data production over decades, including observations, state-of-the-art Earth system models, and advanced data assimilation processes
- Minimal gaps in data collection and minimal service interruptions
- Provide a clearinghouse of data records associated with essential climate variables
- Provide data records with global coverage and policy-relevant spatial-temporal resolution at all governance scales, from nations to individual landowners
- Latency of climate data record production consistent with decision maker time lines

- Decision support tools to enable synthesis assessment and translation of climate data records into societal benefit metrics
- Relentless attention to uncertainty quantification, calibration, and validation of data
- Transparency and reproducibility of observational data, models, and decision support tools and analysis
- Expert scientific interpretation and support for assessors and end users
- Sustained support for ongoing climate research
- Flexibility to respond to lessons learned from research, including coordination with non-operational, research-driven observations

Approaches for implementing a national Climate Information System or climate service in the United States has been a topic of discussion within the scientific community for some years.[149] The USGCRP, World Meteorological Organization (WMO), and Group on Earth Observations are principal leaders in identifying the needs and implementation options for climate services and related observational systems.[150] NOAA has aggressively worked to establish a National Climate Service for the United States.[151,152] To date, the NOAA Climate Service effort has largely focused on reorganizing the agency to align existing NOAA observational systems, data centers, line offices, laboratories, six Regional Climate Centers, and nine Regional Integrated Science and Assessment Centers under a single entity.[153] The US Department of Interior has independently established eight Climate Science Centers[154] that have some overlap with NOAA's regional centers. The NOAA Climate Service concept builds on the resources of the operational National Weather Service and also seeks to incorporate information from the broader community (e.g., USGCRP agencies such as NASA, Department of Interior/USGS, Department of Energy, National Science Foundation, and the United States Department of Agriculture.

Since the limitations in current climate observations and models were described in Chapter 1, the remainder of this section focuses on the other key elements of a climate information system and related processes.

Validation and Uncertainty Quantification

Assessment efforts such as the IPCC process, as introduced in Chapter 1 of this report, provide a critical scientific resource for climate change understanding and assessment as well as for potentially informing economic and political decision-making. A central element of these Assessment Reports is the climate model projections embedded in the Working Group I report(s) and indirectly in the impacts/adaptation and mitigation Working Group II and III reports. These quantitative projections are based on dynamical, multi-component, coupled global climate models (GCMs), now often referred to as Earth system models as they become more comprehensive (e.g., including the carbon cycle). Global observations have long been essential for providing resources for model development and validation (e.g., Tropical Rainfall Measuring Mission (TRMM) precipitation, Earth Radiation Budget Experiment/Clouds and the Earth's Radiant Energy Systems radiation budget, International Satellite Cloud Climatology Project cloud cover, and SSM/I water vapor). In the context of IPCC, such observations are becoming essential as a means for quantifying the uncertainties associated with these climate change projections. In fact, the rather new and informal research and development activities associated with model metrics for GCMs is moving toward more formal development and application of metrics and skill scores which can be, and are being, explored for use for quantitative weighting of an ensemble of model projections.

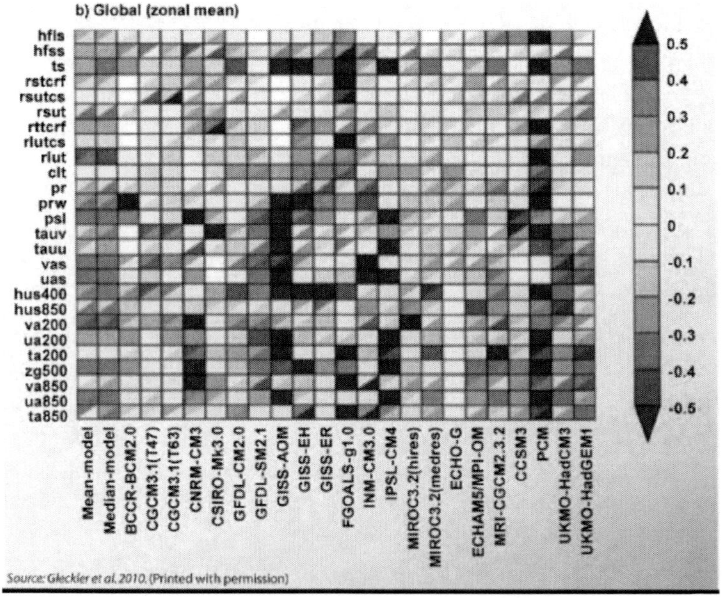

Figure A-3. Portrait diagram display of relative error metrics for 20th century CMIP3 annual cycle, climatology (1980-1999) for zonal mean (with bias removed)[157].

Figures A-3 and A-4 show results from two recent examinations of GCM fidelity based on Coupled Model Intercomparison Project (CMIP) GCM climate simulation archives. These studies are two of the first[155] that attempted to systematically and quantitatively score GCM fidelity across the CMIP archive(s) and across a number of quantities/processes. These attempts, and their further refinements, are crucial for quantitative weighting of future climate projections (e.g., projections from poorer performing models for the observed climate would be given less weight). Figure A-3 shows an example of a portrait diagram from Gleckler et al.[156]

The portrait diagram illustrates GCM fidelity measures for representing the recent climate based on observations and the CMIP3 model archive for the 20th century GCM simulations, and on the utilization of a given metric, in this case, a relative root mean square difference in the annual cycle of the zonal mean of a number of quantities. Note blue colors indicate the model performs better than the typical model in the entire group, and the two left columns indicate that an ensemble-mean of the models performs better than any given single model.

For both of these cases, the observations are based on a number of reanalysis products and a few satellite quantities. The latter includes the satellite data based on Global Precipitation Climatology Project (GPCP)/Climate Prediction Center Merged Analysis of Precipitation for rainfall, International Satellite Cloud Climatology Project for total cloud cover, NASA Water Vapor Project for precipitable water, and Earth Radiation Budget Experiment/Clouds and the Earth's Radiant Energy Systems for shortwave and long wave radiation products. Note however, the wealth of global satellite data products that are going unutilized in such evaluations, including most of the Earth Observation System and A-Train platforms and instruments (e.g., Terra, Aqua, Aura), as well complementary Earth Science platforms such as GRACE, TOPEX/JASON-1/Ocean Surface Topography Mission, and the suite of ocean vector wind products.

Figure A-4 shows results from Reichler and Kim[158] who utilized observation-based metrics to examine and characterize the degree that models as a whole have improved in skill over time. They applied a rather simple metric based on the annual mean climatology and over a number of variables to the suite of model simulations that made up the (~1991) CMIP1, (~ 1998) CMIP2, and (~2005) CMIP3 model archives. In this case, there is a clear illustration that application of a consistent set of metrics to a multigenerational set of models indicates increased model fidelity over the last two decades. As with the Gleckler et al. study, the direct information from satellite observations is extremely limited given the breadth of available observations and only involves, in this case, precipitation, cloud cover, sea ice, sea surface temperature, and the indirect inputs to model-influenced reanalysis quantities.

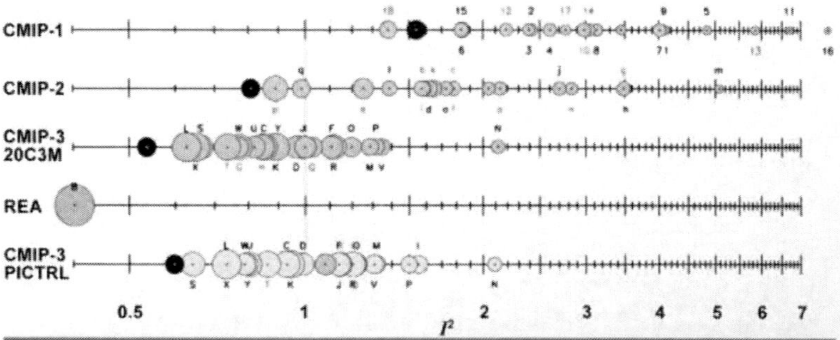

Variable	Domain	Validation data	Period
Sea level pressure	ocean	ICOADS (Woodruff et al. 1987)	1979–99
Air temperature	zonal mean	ERA-40 (Simmons and Gibson 2000)	1979–99
Zonal wind stress	ocean	ICOADS (Woodruff et al. 1987)	1979–99
Meridional wind stress	ocean	ICOADS (Woodruff et al. 1987)	1979–99
2-m air temperature	global	CRU (Jones et al. 1999)	1979–99
Zonal wind	zonal mean	ERA-40 (Simmons and Gibson 2000)	1979–99
Meridional wind	zonal mean	ERA-40 (Simmons and Gibson 2000)	1979–99
Net surface heat flux	ocean	ISCCP (Zhang et al. 2004), OAFLUX (Yu et al. 2004) 1984	(1981)–99
Precipitation	global	CMAP (Xie and Arkin 1998)	1979–99
Specific humidity	zonal mean	ERA-40 (Simmons and Gibson 2000)	1979–99
Snow fraction	land	NSIDC (Armstrong et al. 2005)	1979–99
Sea surface temperature	ocean	GISST (Parker et al. 1995)	1979–99
Sea ice fraction	ocean	GISST (Parker et al. 1995)	1979–99
Sea surface salinity	ocean	NODC (Levitus et al. 1998)	variable

Source: Reichler and Kim 2008.
(c)American Meteorological Society. (Reprinted with permission.)

Figure A-4. Performance index I2 for individual models (circles) and model generations (rows)[159].

The utility of observations, particularly satellite observations, cannot be overstated for their use in model development and evaluation. Figure A-5 shows results of an analysis of 20th century CMIP3 model output adapted from Waliser et al.[160] that illustrates inter-model agreement in long-term annual and global mean values of the four quantities: precipitation, precipitable water, cloud fraction, and integrated cloud ice water path. Notable is that the inter-model agreement between precipitation, precipitable water, and cloud fraction is relatively good. These quantities have relatively long-lived (since late 1980s or before) global satellite records for which model development and evaluation have been able to constantly utilize (e.g., TRMM, GPCP, the Climate Prediction Center's Merged Analysis of Precipitation for rainfall; SSM/I, NASA Water Vapor Project for precipitable water; International Satellite Cloud Climatology Project for cloud fraction). In contrast, the inter- model agreement for the ice water path is quite poor. Even when the greatest outliers are removed, there is still a factor of six between the remaining largest and smallest modeled values. Fortunately, in the case of ice water path, more robust retrievals have become available with the advent of CloudSat, and these large model uncertainties are expected to greatly improve.[161]

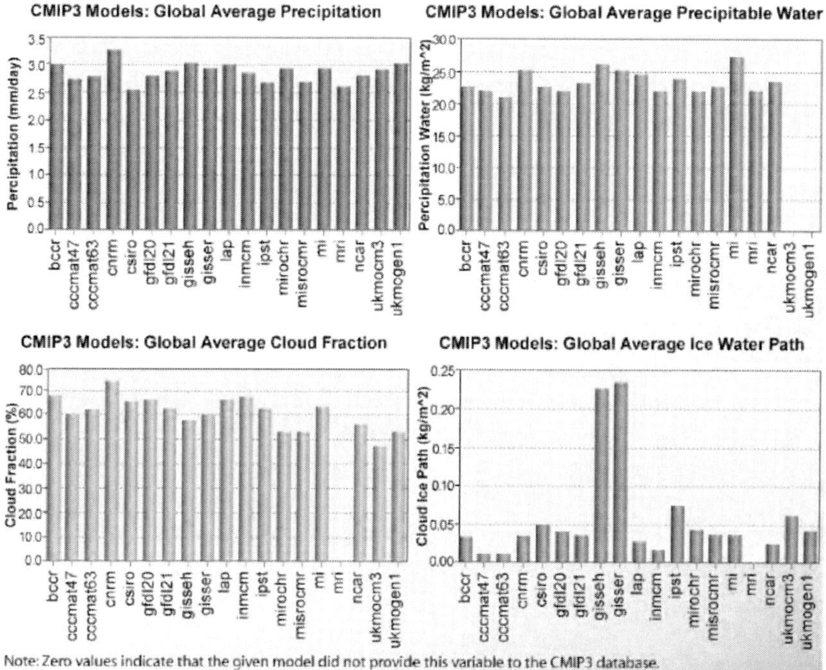

Note: Zero values indicate that the given model did not provide this variable to the CMIP3 database.
Source: Waliser, et al. 2007 (Printed with permission)

Figure A-5. Globally-averaged, annual mean values of hydrological quantities from the 1970-1994 period of the 20th century atmosphere-ocean coupled simulations assessed in the IPCC AR4.

Given the role that GCMs and their simulations and projections play in the IPCC process, not only to the science of climate change but also possibly to decision support associated with some of the most socially- important impacts of global climate change, it is imperative to consider the above finding(s) more earnestly, and to define and take steps to develop and explore the use of more formal metrics and validation procedures in order to gauge the fidelity of the GCMs that contribute to synthesis assessments, such as IPCC.

For example, the numerical weather prediction community routinely uses metrics to provide a uniform and quantitative methodology to evaluate models and their projections and track their improvement over time, with results openly published by the WMO. To date there is no analogue in the climate modeling community. To address this, the Working Group on Numerical Experimentation and the Working Group on Coupled Modeling with sponsorship from WMO's World Climate Research Program have jointly formed a Climate Metrics Panel. From 2010–2011, the task force was active in developing a proposal to start identifying and applying a set of metrics to climate model results with a particular near-term objective, CMIP5. The main questions that are motivating the use of routine climate metrics: Are climate models improving? If they are improving, then how rapid is the improvement? Which are more realistic?

Additionally, it is hoped that a diversity of routine metrics will provide useful summaries of overall model performance. Such metrics will in turn be useful to explore as a means to objectively weight the ensembles of climate projections provided as part of CMIP5 (e.g., for one greenhouse gas scenario, approximately twenty GCMs and possibly five ensemble members ~ one hundred). Note that similar considerations and metrics need to be developed for regional climate models that are routinely used for dynamical downscaling to scales more relevant to local and regional decision support.

Despite a growing body of models, two broad issues remain in their effective application to climate risk assessment: 1) model validation and scoring are needed so that when assessors evaluate the outputs of multiple models offering projections for a given region, they have some objective way of weeding out and/or weighting projections; and, 2) the lack of capacity of decision makers to effectively interpret and use the information.

The application of global satellite products to model development and evaluation leads to models that have more fidelity at representing the observed climate record.

The development and application of observation-based metrics provide a formal and uniform methodology to evaluate models, track their improvement over time, and potentially provide objective, skill-dependent weighting of their predictions and projections. This statement is already clearly demonstrated by the numerical weather prediction community.

A significant portion of NASA and other agency satellite data is being underutilized for model development and evaluation and, particularly, for use as observation-based climate model metrics.

Many processes within climate models could be better understood and modeled if new observational resources were made available.

Climate Data Record Production

In order to produce operational climate data records, significant effort is required to accumulate data associated with ECVs, rigorously quantify uncertainties, apply calibration corrections, and transform time-series of data into trends. Additionally, calibrated data can be rectified and presented as geolocated maps indicating the spatial and temporal distributions and evolutions of selected parameters.

To support transparency and reproducibility, these observational data sets must be accompanied by extensive metadata covering details on the sensing instrumentation, algorithms employed to retrieve geophysical parameters from the raw data, calibration curves, etc. Maintaining an archive and clearinghouse for these data including expert interpretation and support for users is a major component of any climate information system. NOAA's existing National Climatic Data Center, National Oceanic Data Center, and National Geophysical Data Center are several such assets and form the core of the NOAA Climate Service.

NASA's Earth Observing System Data and Information System manages and distributes data products through the Distributed Active Archive Centers[162] spanning twenty-eight categories of ECVs covering the atmosphere, oceans, and land surface.

USGS's Earth Observations and Science Center[163] provides access to climate-relevant information including carbon cycle, drought, vegetation monitoring, land cover/land use, fire, wildlife impacts, and topography.

In terms of model outputs, in addition to the above outlets, various organizations and programs such as the IPCC[164] and the Program for Climate Model Diagnosis and Intercomparison[165] offer the results of individual and ensemble model simulations and projections.

In terms of the above data resources, data measurement, and archiving strategies, few efforts have systematically optimized observation formats, archives, and dissemination strategies for the specific use of climate model evaluation and improvement.

Decision Support Tools

The availability of robust climate data records, trends, and spatially resolved distributions of essential climate variables alone is not sufficient to support accurate assessment of climate change risks and decision- making. The concept of decision support tools recognizes the need for applications that help bridge the gap between climate science and societal benefit areas.

Climate decision support tools are funded and produced by government agencies (both research and operational entities), think- tanks, academia, aid organizations, and private industry.

In the framework of this report, decision support tools are treated as part of climate information systems which are arguably the best home for such capabilities. The following examples offer a noncomprehensive illustration of the scope and functions of several existing climate-relevant decision support tools.

The National Integrated Drought Information System[166] is managed by NOAA with support from other agencies and designed for the United States only to:

- Provide early warning about emerging and anticipated droughts
- Assimilate and quality control data about droughts and models
- Provide information about risk and impact of droughts to different agencies and stakeholders
- Provide information about past droughts for comparison and to understand current conditions
- Explain how to plan for, and manage, the impacts of droughts
- Provide a forum for different stakeholders to discuss drought- related issues

The Famine Early Warning System Network (FEWS NET) is a USAID-funded activity implemented by a partnership between USGS, NASA, NOAA, US Department of Agriculture, and Chemonics International, Inc. FEWS NET collaborates with international, regional, and national partners to provide timely and rigorous early warning and vulnerability information on emerging and evolving food security issues. FEWS NET has regional centers in Central Asia; the Caribbean; Central America; Eastern, Southern, and Western Arica; and the Middle East.[167]

The Foreign Agricultural Service's Crop Explorer[168] features near-real-time global crop condition information based on the satellite imagery and weather data processed by the Production Estimates and Crop Assessment

Division. The primary mission of this division is to produce the most objective and accurate assessment of the global agricultural production outlook and the conditions affecting food security in the world.

Regional analysts use a Geographic Information System to collect market intelligence and forecast reliable global production numbers for grains, oil seeds, and cotton. Thematic maps of major crop growing regions are updated every ten days to depict the latest statistics pertaining to vegetative vigor, precipitation, temperature, and soil moisture. Time- series charts depict current and historical growing season data for specific agro-meteorological zones. Regional crop calendars and crop area maps are also available for selected regions.

The Regional Visualization and Monitoring System initiative, a USAID-funded activity implemented by NASA in collaboration with regional partners in Central America, Africa, and Central Asia, integrates satellite observations, ground-based data, and forecast models to monitor and forecast environmental changes and to improve response to natural disasters.[169]

One feature common in many climate decision support tools is visualization, particularly the application of Geographic Information System techniques. The concept of risk mapping is frequently employed in focusing vulnerability assets in regions such as Africa.[170] In fact, most of the briefings given to this task force over the course of the study presented a large number of maps constructed with Geographic Information System tools illustrating climate impacts and vulnerabilities of various regions. While the motivation for spatially resolved visualization is understandable, some caution in their interpretation is warranted.

Most Geographic Information System tools employ one or more observational data sets and/or model outputs to visualization. This information is not necessarily applied consistently (e.g., as layers or as merged data sets) and significant errors can exist, both in terms of registering data to a given grid and in terms of differences between underlying data sets.

Methods for representing and interpreting uncertainty or errors in Geographic Information System products are by no means consistent or widespread. Hence, there exists a risk of mixing gray data, data lacking verifiable pedigree or quantified uncertainties, with high reliability information.

The potential for policy getting ahead of the science is a topic of some concern within the scientific community. At the same time, it is recognized that decision makers cannot wait for zero uncertainty.

Findings on Climate Record Data Production

Despite the wealth of information available from multiple organizations there currently is no single clearing-house for climate data records and associated model outputs, nor identification or arbitration of discrepancies between them.

As with climate information systems writ large, there is no standard set of decision support tools or even standards for how they are designed, implemented, and used, nor are there conventions on who produces them.

There is a potential for well-presented but inaccurate gray data associated with Geographic Information System products to receive higher weighting by risk assessors than other, less intuitive, but more accurate information.

Information products used in climate assessment processes are often not accompanied by clearly marked quantitative uncertainty estimates to ensure appropriate weighting by decision makers.

Overarching Barriers for Climate Information Systems

In addition to issues with specific elements of climate information systems, there are several overarching barriers to deploying a robust, operational capability principally associated with:

- Funding priorities place an inordinate weight on control over knowledge
- US climate services focused on the United States, not international needs
- Inactive research to operations pathway for climate information
- Limited capacity in developing countries for information systems

Funding for climate information systems and synthesis assessment (knowledge), receives a small fraction of funding available for response actions (control). For example, only 0.4 percent of the African Development Bank's current adaptation budget ($45 million out of $5.9 billion) is allocated to knowledge and competency building.[171]

In the United States, the funding spread across NASA, NOAA, USGS, Department of Energy, National Science Foundation, Department of Agriculture, and other agencies for 2007-2009 allocated to research associated with adaption and mitigation was approximately 25 percent, or $300 million per year, of the ~$1.2 billion total USGCRP budget.[172]

However, this again represents the climate information (knowledge) component of climate change risk management, not the response (control) aspect. Risk managers have noted the need to create "well-designed and adequately resourced feedback loops to effectively incorporate new data and advancements in scientific understanding and support continual refinement and validation of analyses, impact projections, and effective response mechanisms."[173]

This finding and the need to increase efforts to offer a broader global focus relevant to the needs of the defense, diplomacy, and development agencies was also noted in the recent progress report of the US Interagency Climate Change Adaptation Task Force.[174]

A research to operations process has been employed for decades to develop and transfer scientific information systems from an exploratory science mode into operational capabilities.

Research agencies such as NASA, the national laboratories, and their partners in academia serve as pathfinders for research and development of instrumentation, data products, and models that are gradually (over a decade or more for a given ECV) transferred into an operational environment by agencies such as NOAA or USGS driven by strategic planning.

It is not clear if the US government is on such a path for a climate information system. The practical planning window for these agencies is roughly ten years, and the current horizon regarding climate observations is roughly 2020.

A fair concern often expressed in considering operational climate services is the need to remain flexible to discovery. The key point here is recognition of the need to *start* such a process.

Currently, efforts to address this gap are limited to projects funded by USAID, such as NASA's Regional Visualization and Monitoring System. These pilot efforts offer good test particles for the deployment of climate information products, technical infrastructure, and training in developing countries but are still limited in the scope, funding, and breadth of US science and technical capability employed.

The Interagency Climate Change Adaptation Task Force recognized the same need and made the following recommendation: "Develop a Government-wide strategy to support multilateral and bilateral adaptation activities and integrate adaptation into relevant U.S. foreign assistance programs."[175]

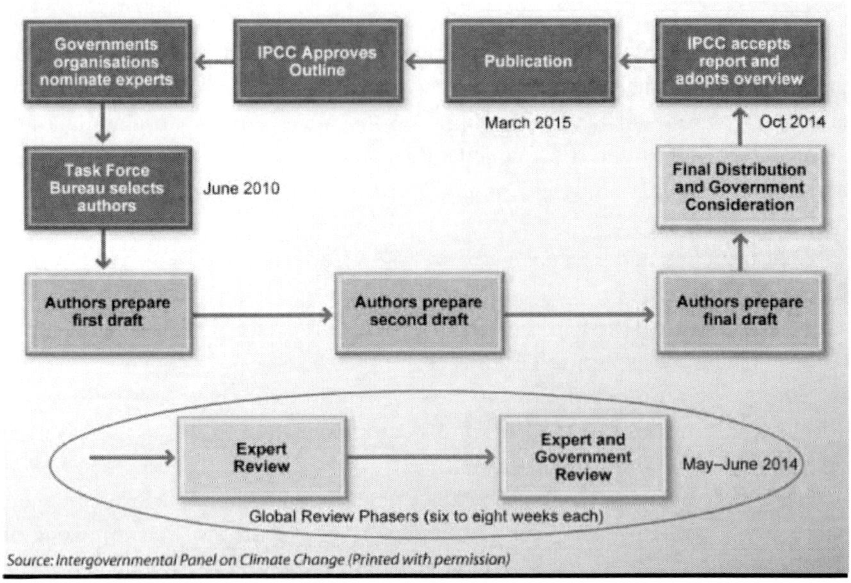

Figure A-6. IPCC Fifth Assessment process and timeline.

Synthesis Assessment

The outputs of climate information systems and other information systems must be integrated and treated to synthesis assessment in order to translate biogeophysical data into metrics relevant to water, food, shelter, energy, security, health, and other societal impact areas. This requires the application of an interdisciplinary approach involving the physical sciences, social sciences, economics, and policy. Beyond studying current and past climate data and other data sets, synthesis assessment benefits from defining impact and response scenarios and running simulations to evaluate potential outcomes and risks.

The IPCC offers perhaps the best known example of synthesis assessment for climate change. Four IPCC assessment reports have been released to date: 1990, 1995, 2001, and 2007. The fifth assessment is currently underway and scheduled for publication in early 2015, with a goal of releasing an advanced copy by late 2014.

The process and time line for generating the IPCC synthesis assessments are illustrated in Figure A-6. Unlike an operational climate change risk management framework in which climate and other information systems produce sustained, operational inputs to a synthesis assessment function, the IPCC process is strictly limited to considering only information available in the peer-reviewed scientific literature at given cut-off dates. This is intended to maximize the credibility of the foundational information applied in the assessment. This requirement is both necessary and presents a risk for decision makers. On the one hand, a deliberative process to ensure that the climate information presented is scientifically robust is not amenable to rushing. The peer-review process is important. On the other hand, there is significant latency and perhaps major gaps in knowledge associated with a process strictly limited to peer- reviewed journal articles. Additionally, as illustrated, the IPCC assessment report must be reviewed and approved by a large number of governments prior to release. The risk here is that vital and relevant climate information can be discarded, leaving with the end-result being a diluted version of the complete story.

The USGCRP is required under the Global Change Research Act of 1990 to provide a National Climate Assessment report every four years that includes an analysis of the effects of global change on natural environment, agriculture, energy production and use, land and water resources, transportation, human health and welfare, human social systems, and biological diversity.[176] It also analyzes current trends in global change, both human-induced and natural, and projects major trends for the subsequent twenty-five to one hundred years. So far, two such national assessments have been produced, one in 2000, the other in 2009. The next assessment is due in 2013 and will differ from the previous assessments by being "more focused on evaluating the Nation's progress in adaptation and mitigation building a long-term, consistent process for evaluation of climate-related risks and opportunities, including a national set of indicators of change; providing information that supports decision-making processes within regions and sectors of the United States; and evaluating the current state of scientific knowledge relative to climate impacts and trends."[177] Unlike the IPCC assessments, the USGCRP national assessment, other than noting global climate trends, is limited in scope to the United States. While the national assessment allows a deeper focus on regional and local issues in United States than the IPCC assessment, to a large extent, it does not address international issues.

Recently, synthesis assessments focused on specific priority areas have been conducted by organizations in the United States. Examples include the Environmental Indications and Warnings (EIW) Project implemented by the CIA Center on Climate Change and National Security, conducted under the auspices of the Measurements of Earth Data for Environmental Analysis program. The EIW seeks to monitor environmental stresses influenced by climate, using measurements and modeling of security indicators descriptive of the coupled human-environmental system. Initially, the focus of the EIW project is on freshwater availability, ultimately offering global coverage and sub-national resolution with monthly updates on forward projections. The EIW assessment includes a synthesis of climatic stresses (specific anomalies such as temperature, precipitation, soil moisture, and river flow), environmental factors (bio/geo/hydrosphere), socioeconomic and political factors to identify and risk-rate water resource hot spots.

Another example of a focused assessment includes a risk analysis by the Department of Energy's Sandia National Laboratories that assembled IPCC climate model ensemble projections for temperature and precipitation, used a hydrologic model to map those projections to future water availability and water production, then applied a macroeconomic model to estimate the potential economic consequences for the United States as a whole and individual states for the period 2010-2050.[178]

The need for improved collection and integration of multiple information types (climate, economic, security, demographics, health, and other indicators) cannot be overstated. Figure A-7 illustrates the results of a synthesis analysis by Meier et al. that combined climatic data (precipitation) and other information (human deaths and livestock losses) attributed to increasing conflict in the form of cattle raids in Uganda during the period of peak water stress before the start of the wet season.[179]

This need is supported by the following recommendation from the National Research Council report, *Informing Decisions in a Changing Climate*. The "federal government [should] 'expand and maintain national observation systems to provide information needed for climate decision support. These systems should link existing data on physical, ecological, social, economic, and health variables to each other and develop new data and key indicators as needed' for estimating climate change vulnerabilities and informing responses intended to limit and adapt to climate change."[180]

A complete and robust climate change risk management framework requires a reliable, sustained, and timely synthesis assessment component. The Interagency Climate Change Adaptation Task Force[181] recognized this point and offers the following specific recommendations to:

- Develop scenarios within a range of climate change outcomes at spatial and temporal scales necessary to inform impact assessment and adaptive action
- Create user-friendly methods for assessing climate impacts, vulnerability, and risk, including models and tools to assess the

environmental, social, and economic outcomes of alternative adaptation actions

- Provide guidance on the use and suitability of downscaled global climate model outputs
- Conduct frequent updates of regional characterizations and assessments of climate, including climate-driven variables (e.g., stream flow, flood, and drought)
- Expand research on relevant social and behavioral sciences to improve understanding of human responses to change
- Identify the social and ecological tipping points and thresholds (beyond which change is sudden and potentially irreversible) to help guide decisions regarding intervention and planning
- Develop methods and processes for identifying, defining, and managing for extremes, including low-probability, high-impact events
- Analyze climate change impacts in the context of multiple stressors and interacting systems (e.g., interactions of climate and air quality on human health in metropolitan areas)

It remains to be seen to what extent these recommendations will be implemented such that a synthesis capability, meeting the needs of the Department of Defense, will be available in the future.

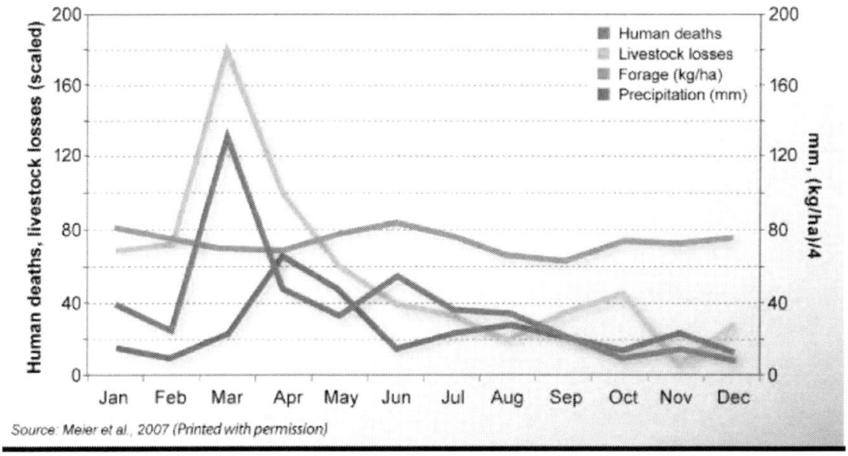

Figure A-7. Conflict impact and precipitation levels for Ugandan Karamojan.

Decision-Making and Integration

Effective climate change risk management depends both on the quality of the available knowledge (climate information systems and synthesis assessment process) and the process used to manage control and responses (decision-making). Climate change risk management introduces challenges that build on those encountered in more traditional decision- making. These challenges include fundamental uncertainties in climate information, the need to consider many low-likelihood, high impact risks, and the complexity and risk of competing action associated with asynchronous responses from a large number of management entities both in the United States and internationally.

Given the significant gaps in scientific understanding of climate change, it is likely that policy-relevant climate information systems may take a decade or more to deploy. In the meantime, risk managers are confronted with the need to move forward in the face of uncertainty.

As noted in a 2010 report by Sandia National Laboratories: "An imprecise prediction can be useful for comparing options to address a significant problem if we assume that such a prediction adequately defines the future relative to the choices to be made and, more importantly, represents a mutually agreed upon basis from which stakeholders can debate alternatives on common ground."[182] The treatment of low probability, very high impact risks whose probability density functions exhibit long tails is an area of concern for risk managers. Climate change presents a potentially large phase space of scenarios with the potential for far greater and more widespread impacts with arguably comparable likelihoods but with relatively poorly understood response options (Appendix B). This presents a challenge for risk management processes that probably warrant further study to enumerate specific gaps and mitigations.

The other barrier to effective decision-making in this arena is the size of the potential response space and the number of organizations involved in evaluating risk and taking action. In the United States alone, there are approximately ten agencies associated primarily with the generation of knowledge/information and at least fifteen associated with control/action, plus a few that overlap. The present lack of any overarching framework for coordination and integration represents a risk of asynchronous actions. Some potential barriers include:

- Challenges for interagency cooperation and coordination
- Interagency coordinating organizations such as the President's Office of Science and Technology Policy, USGCRP, USGEO, and the Interagency Adaptation Task Force not acting as delivering

organizations, many relying on contributed efforts by many agencies which is an inefficient process

- Intra-agency issues such as conflicting visions within large departments and agencies, competition for funding, and a wariness of unfunded or underfunded mandates
- Lack of direct cabinet level representation from agencies responsible for delivering climate information systems (e.g., NASA, NOAA, USGS)
- Lack of an empowered integrated decision-making body; no climate equivalent of the National Security Council

The interagency adaptation working group had a similar finding:

Many programs across the Federal Government produce science that informs and supports climate change adaptation decision-making. Many of these efforts occur through the agencies of the USGCRP, while others have emerged in resource management or community development programs through agencies that have not historically focused on climate change. Currently, most of these activities are occurring independently of one another, leading to gaps and redundancies. These efforts would benefit from enhanced coordination on science at the Federal level, through agencies working together more closely to leverage existing capabilities. Coordination would help federally sponsored science identify, understand, and meet the needs of decision makers implementing adaptation strategies on the ground. The new Adaptation Science and Research Element within the USGCRP should develop a 'roadmap' that identifies existing adaptation science and service capabilities and gaps across Federally-sponsored programs. [183]

Findings on Overarching Barriers for Climate Information Systems

Currently, climate observations, regional-scale models, decision- support tools, and synthesis assessments by the USGCRP and particularly by the NOAA and USGS contributions are primarily focused on the United States rather than an international perspective.

There remain many uncertainties in basic climate process understanding that present barriers to specifying an optimal, sustained information system with confidence, suggesting that an iterative or spiral development approach will be required.

There is limited scientific and technical capacity in the developing world to develop and maintain climate information systems.

The IPCC assessment process, while rigorous and comprehensive, suffers from latency and potentially, from the diluting influence of multinational politics on the ultimate findings.

Absent a focused effort to leverage and augment the IPCC and USGCRP assessment efforts, there exists a significant risk the synthesis assessment needs will not be met.

There is not a systematic effort to quantify and recognize the uncertainties embodied in climate and other information resources and the range of potential response options; this effort must be done consistently across a wide-range of decision-making organizations in such a way as to provide a level playing field for risk assessment and to ensure responses are harmonized.

Summary

Various organizations in the United States are working to address the general need for improved information to support climate change risk management. The USGCRP is currently undergoing a restructuring activity intended to improve its capability to support the climate change response efforts.[184] The Climate Change Adaptation Interagency Task Force has broad participation of federal agencies and has identified numerous areas for improvement, including some relevant to the needs of DOD and other agencies concerned with international threats.[185] NOAA and other individual agencies continue efforts to improve the relevance of their information to climate change risk management and decision-making.

Nevertheless, neither the needs of DOD for climate information systems and assessment processes, nor the gaps relative to current capabilities and planned future capabilities have been rigorously studied to date. Likewise, many of the overarching barriers identified here, and in other studies, are unlikely to be resolved without more attention.

APPENDIX B. SPECIAL TOPICS

This appendix elaborates on several wild card topics, each of which have the potential for placing additional demands on climate information systems and risk assessment processes beyond those discussed earlier.

Tipping Points

The term "tipping point" commonly refers to a critical threshold at which a tiny perturbation can qualitatively alter the state or development of a system; the term "tipping element" has been used to describe large- scale components of the Earth system that may pass a tipping point.[186] Figure B-1 illustrates potential large-scale tipping elements, ranging from changes in ENSO amplitude and frequency, to rapid loss of Arctic sea ice or key ice sheets (Greenland and Antarctica), to major biosphere perturbations in the Sahara, Amazon, and Boreal regions, each with significant global reach through teleconnections and potential instability due to positive feedbacks.

This topic has been recognized as a wild card for climate risk management. The National Research Council's (NRC) "America's Climate Choices" study found that:

> ...rather than smooth and gradual climate shifts, there is the potential that the Earth system could cross tipping points or thresholds that result in abrupt changes. Some of the greatest risks posed by climate change are associated with these abrupt changes and other climate "surprises" (unexpected changes or impacts), yet the likelihood of such events is not well known. Moreover, there has been comparatively little research on the impacts that might be associated with "extreme" climate change—for example, the impacts that could be expected if global temperatures rise by 10 °F (6 °C) or more over the next century.[187]

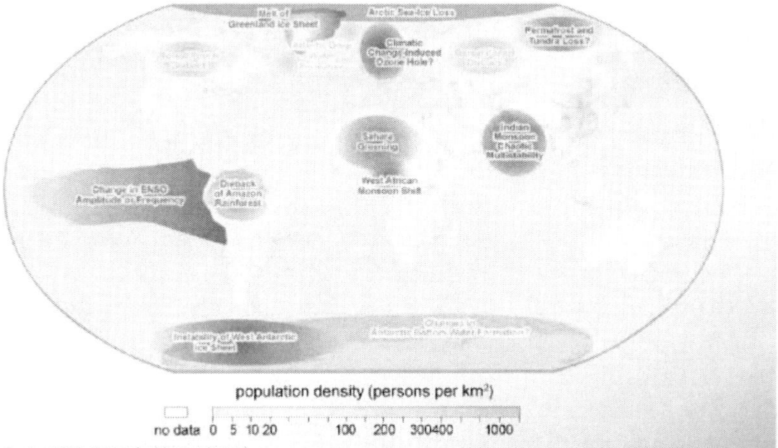

Source: Lenton 2008. (Printed with permission)

Figure B-1. Potential tipping elements and their approximate geographic scope.

The possibility for a given parameter to undergo a tipping point event, such that the climate system undergoes a qualitative shift into a new stability regime, is illustrated in Figure B-2, where, as described by Lenton et al. the potential wells represent stable attractors, and the ball, the state of the system. Under gradual anthropogenic forcing (progressing from dark to light blue potential), the right potential well becomes shallower and finally vanishes (threshold), causing the ball to abruptly roll to the left. The curvature of the well is inversely proportional to the system's response time to small perturbations.[188]

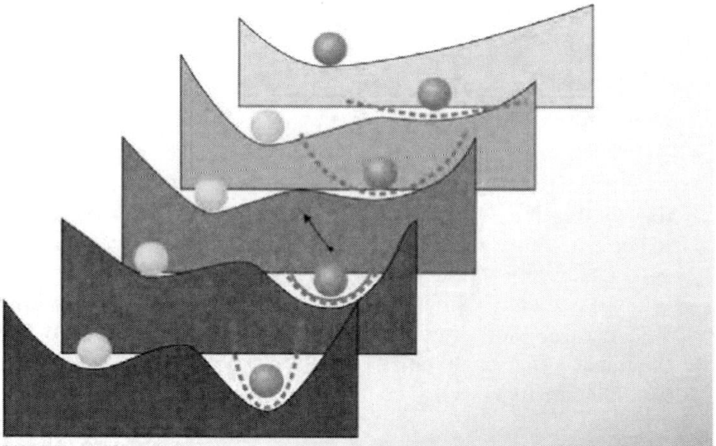

Source: Lenton et al. 2007. (Printed with permission)

Figure B-2. Concept view of a tipping point or threshold crossing event leading to an abrupt change in climate state and future stability.

The impact of such sudden transitions is twofold. One, the shift could place the climate system in a qualitatively different stability regime. For example, analysis of ice core data suggests such a tipping point event about twelve thousand years ago precipitated the rapid end of the Younger Dryas, an approximately thirty-year transition period of extreme climate variability, characterized by large changes in temperature and precipitation on timescales as short as three years.[189] There continues to be debate as to whether the Younger Dryas encouraged or impacted the advent of agriculture. It is not clear whether modern dependence on large- scale agriculture would be resilient to the rapid transients in temperature and precipitation observed during the Younger Dryas termination event.

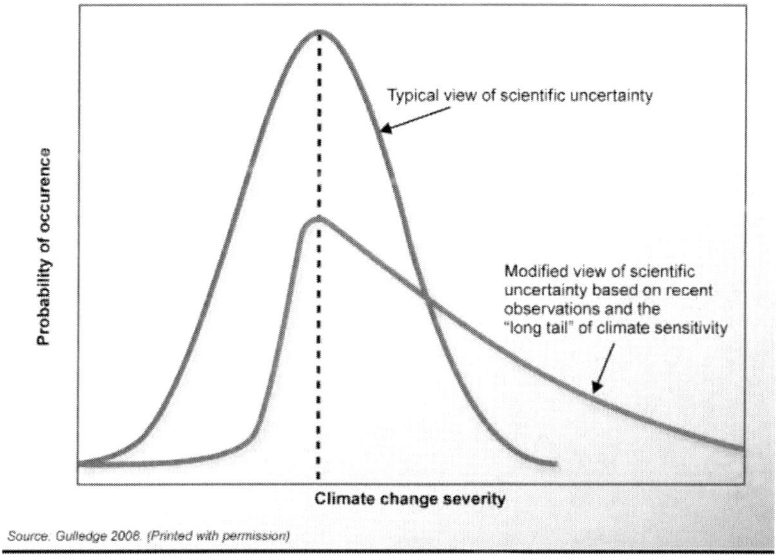

Figure B-3. Tipping elements in the Earth's climate system.

The duration of such climate epochs also illustrates the second factor of reversibility. While the 1300-year duration of the Younger Dryas might not be irreversible in geologic terms, it would certainly be considered irreversible on societal timescales. With regards to irreversibility, the NRC *America's Climate Choices* study found that "There is general scientific consensus that the Arctic, which is systematically losing summer sea ice thickness and extent on an annual basis, is expected to become permanently ice-free during summers by the middle of the 21st century, regardless of how future emissions change. This change to an ice-free summer Arctic is expected, in part, because of the positive feedback between warming and sea ice melting."[190]

Another foundational issue in assessing the risk of climate tipping point scenarios is the long tail challenge. Figure B-3 illustrates the long tail distribution of potential climate change outcomes, driven by the skewed distribution of probabilities associated with the combined uncertainty in future anthropogenic greenhouse gas emissions and climate sensitivity to that forcing.[191] The fact that the red curve overlays the traditional blue curve suggests the potential of underestimating the true likelihood of worst-case climate change scenarios. This could lead to neglecting observational data collection and modeling for such conditions as well as evaluation of the response options and development of contingency plans.

Finding

Compared to more gradual climate change scenarios, the potential for tipping point events presents additional needs for climate observations (monitoring for early warning signs), models (non-linear processes), and risk assessment processes (managing longtail threats). These are needs that may not receive the priority they deserve.

Geoengineering

The full consequences of future climate change are not yet fully understood. A prevailing view contends that any mean surface warming above about 2°C from pre-industrial times will be dangerous, producing serious negative consequences for humans and natural systems. However, that number represents only one slice through the range of climate sensitivities. The resulting societal impacts are also quite uncertain. The safest and most obvious method of moderating such climate change is to take early and effective mitigation action to reduce emissions of greenhouse gases. However, global efforts to reduce these emissions have not yet been successful, and there is no evidence that the proposed reductions required to avoid reaching the potentially dangerous climate change will be achieved in the near or medium term future.

Additionally, a serious threat multiplier for climate change impacts is the persistence of CO_2 and other long-lived greenhouse gases in the atmosphere, many of which decay more slowly than long-lived radioactive isotopes of fission products from nuclear power plants (Figure B-4). The atmospheric decay times in the figure are basedon a simple exponential decay model for Cesium-137 and use the Bern Carbon model for CO_2 decay. Both start with a unit impulse (normalized to 1.0) at time zero followed by decay per those models for 1,000 years. This irreversibility presents the threat that climate changes, much larger than currently predicted, could persist for many centuries (Figure B-5).

Given our inability to reach agreements to reduce emissions to mitigate climate change, and given the long-term consequences of adding greenhouse gases into the atmosphere today, a number of climate intervention concepts have been proposed. These concepts, referred to as geoengineering, can be divided into two broad classes: 1) Solar Radiation Management techniques that seek to increase the amount of the solar radiation reflected back into space thus increasing the Earth's albedo by a small percentage to offset the effects of increased greenhouse gases; and, 2) Carbon dioxide removal techniques which aim to remove CO_2 from the atmosphere. (Figures B-4 and B-5)[192]

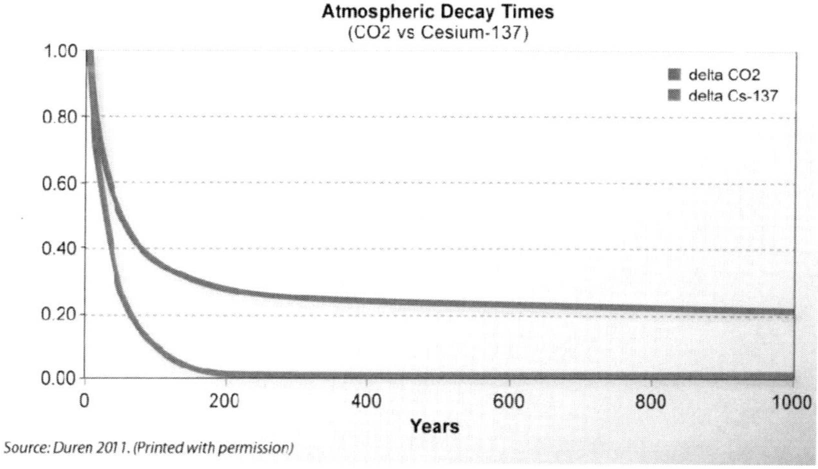

Figure B-4. Atmospheric decay times (CO_2 versus Cesium-137).

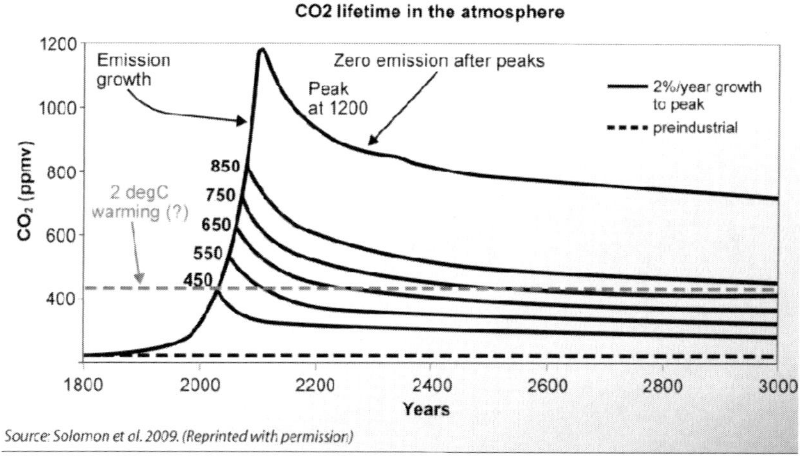

Figure B-5. CO_2 lifetime in the atmosphere.

While such schemes have been discussed in scientific literature for decades, recent studies have focused considerable serious attention towards assessing the potential rewards and risks, including scientific, technical, governance, and sociopolitical issues. This includes the Royal Society's

Geoengineering the Climate report[193] and the NRC's *Advancing the Science of Climate Change* report.[194] In 2010, the US Government Accountability Office conducted a study in response to a request from Congress to assess the state of serious scientific research in the United States, "Climate Change: A Coordinated Strategy Could Focus Federal Geoengineering Research and Inform Governance Efforts."[195]

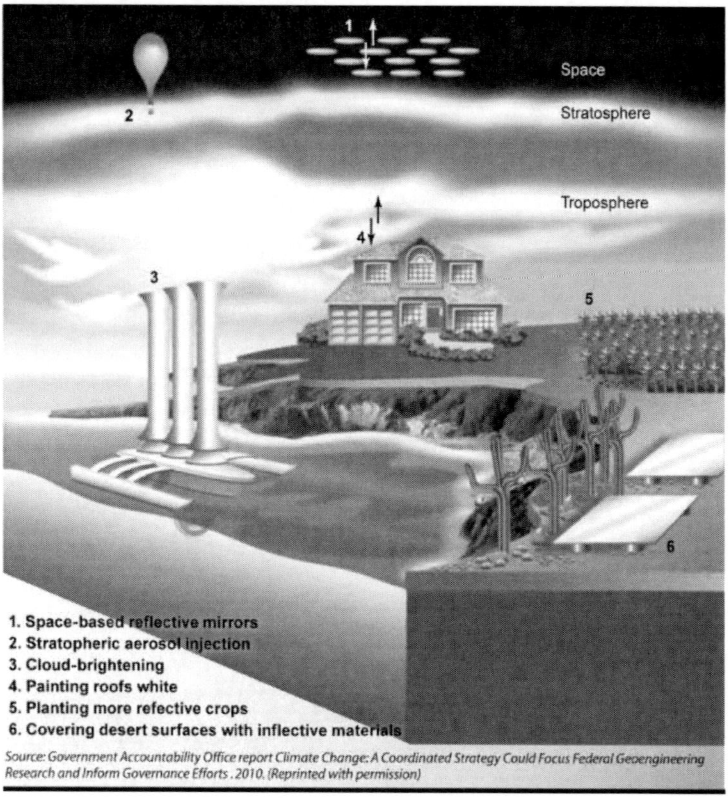

Figure B-6. Carbon dioxide removal geoengineering concepts.

The general conclusion of these reports is that while most of the proposed geoengineering options are considered technically impractical or cost-prohibitive, some, such as stratospheric aerosol injection and boundary-layer marine cloud-seeding, are potentially feasible. These reports also conclude that the potential for unintended consequences for all geoengineering options has not yet received significant study. These reports find that a significant research program would be a necessary precursor to any sub-scale field experiments or full-scale deployment of geoengineering. For example, the Government Accountability Office "recommends that within the Executive Office of the President, the appropriate entities, such as the Office of Science and Technology Policy, establish a clear strategy for geoengineering research in the context of the federal response to climate change to ensure a coordinated federal approach."[196] To date there is no such research program in the United States despite related work in conventional carbon capture and sequestration by the Department of Energy.

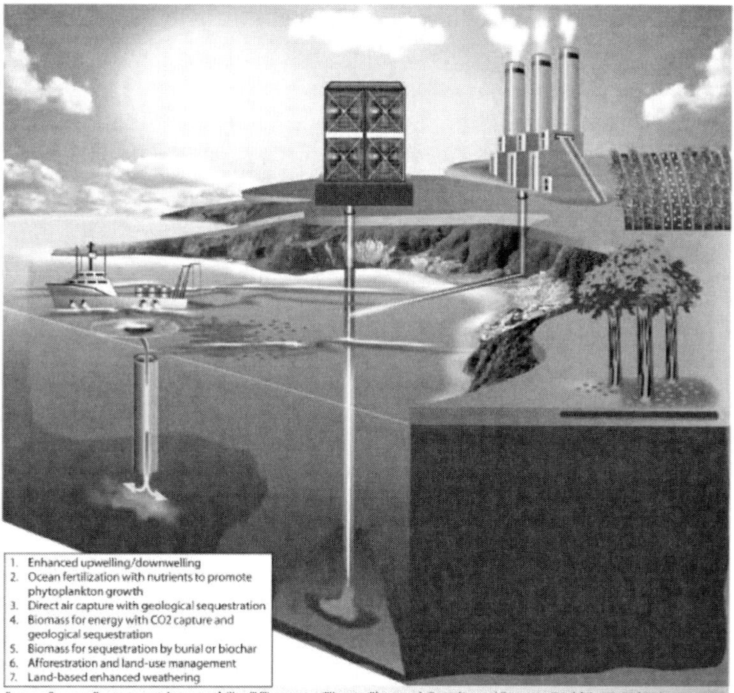

Source: Source: Government Accountability Office report Climate Change: A Coordinated Strategy Could Focus Federal Geoengineering Research and Inform Governance Efforts. 2010. (Printed with permission)

Figure B-7. Solar radiation management geoengineering concepts.

There is a significant potential for unilateral geoengineering activity, both sub-scale experiments and full-scale deployment. For example, for several years China has demonstrated a propensity to attempt modifying weather in Beijing and other areas. Given the global scale teleconnections in the climate system, attempts to modify climate on regional scales has the potential for significant and unpredictable consequences in other parts of the world. Hence, efforts by one nation-state to improve their local conditions could negatively impact a neighboring or remote nation-state.

Solar Radiation Management geoengineering concepts indicate that corporations are already engaging in geoengineering research, in some cases leading to conflict. This threat is compounded given that the technology necessary to implement some of the geoengineering options could be trivial compared to developing nuclear weapons (Figure B-6). Geoengineering nonproliferation may become a real threat over the coming decades.

This risk suggests a potential need for a US research activity focused on improving process understanding in areas that otherwise might not receive prioritization under general climate change science, for example, stratospheric sulphur cycle and impacts of continuous stratospheric aerosol injection on thc ozone layer. It could also drive the need for sustained monitoring systems to detect evidence of unilateral geoengineering activity with precisions and spatio-temporal resolution beyond that required for traditional climate science (Figure B-7).

Glossary

AFRICOM	United States Africa Command
AOR	area of responsibility
AR4	4th Assessment Report (of the IPCC)
C	carbon
°C	Celcius
C4MIP	Coupled Carbon-Climate Cycle Model Intercomparison Project
CCAPS	Climate Change and African Political Stability (program at Strauss Center funded by DOD's Minerva Initiative)
CEOS	Committee on Earth Observation Satellites
CH_4	methane
CIA	Central Intelligence Agency
Cm	centimeter
CMIP	Coupled Model Intercomparison Project
CMIP5	5th Coupled Model Intercomparison Project
CO_2	carbon dioxide
COE	Center of Excellence for Disaster Management and Humanitarian Assistance (at US Pacific Command)

CORDEX	Coordinated Regional Climate Downscaling Experiment
DEIC	DOD Environmental International Cooperation Program
DESDynI	Deformation, Ecosystem Structure and Dynamics of Ice
DOD	Department of Defense
DOS	Department of State
DSB	Defense Science Board
ECMWF	European Center for Medium-Range Weather Forecasts
ECV	Essential Climate Variables
EIW	Environmental Indications and Warnings
ENSEMBLES	A common ensemble climate forecast system used to construct integrated scenarios of future climate change, including both non-intervention and stabilization scenarios
ENSO	El Niño Southern Oscillation
FEWS NET	Famine Early Warning System Network
FI	fossil intensive
FO	follow-on
GCM	Global Climate Models
GDP	Gross Domestic Product
GEMS	Global Environment Monitoring System
GPCP	Global Precipitation Climatology Project
GRACE	Gravity Recovery and Climate Change Experiment
Gt	gigation
H2O	water vapor
ICESat-2	Ice, Cloud, and Land Elevation Satellite-2 (This is the second-generation of the orbiting laser altimeter ICESat scheduled for launch in early 2016.)
IPCC	Intergovernmental Panel on Climate Change
JASON-1	Oceanography mission to monitor global ocean circulation, study the ties between the oceans and atmosphere, improve global climate forecasts and predictions, and monitor events such as El Niño conditions and ocean eddies
Km	kilometer
N_2O	nitrous oxide
NARCCAP	North American Regional Climate Change Assessment Program
NASA	National Aeronautics and Space Administration
NCEP	National Centers for Environmental Prediction

NGO	nongovernmental organization
NOAA	National Oceanic and Atmospheric Administration
NRC	National Research Council
NWP	Numerical Weather Prediction
O3	ozone
OI	optimum interpolation
OSD	Office of the Secretary of Defense
OUSD(AT&L)	Office of the Under Secretary of Defense for Acquisition, Technology and Logistics
OUSD(P)	Office of the Under Secretary of Defense for Policy
PRUDENCE	Prediction of Regional Scenarios and Uncertainties for Defining European Climate Change Risks and Effects
RANET	Radio and Internet for the communication of hydro-meteorological information for rural development
RCM	Regional Climate Models
SMAP	Soil Moisture Active Passive
SOUTHCOM	United States Southern Command
SRES	Special Report on Emissions Scenarios
SSM/I	special sensor microwave/imager
SST	sea surface temperature
SWOT	Surface Water and Ocean Topography
TOPEX	TOPEX/Poseidon satellite
TRMM	Tropical Rainfall Measuring Mission
UNEP	United Nations Environmental Program
USACE	US Army Corps of Engineers
USAID	United States Agency for International Development
USGCRP	United States Global Change Research Program
USGEO	US Group on Earth Observations
USGS	US Geological Survey
WMO	World Meteorological Organization

End Notes

[1] Strauss Center's program on Climate Change and African Political Stability Policy Brief. February 2011. Page 5. Accessed at: http://ccaps.strausscenter.org/system/ research_items/ pdfs/43/original.pdf71299598361. This material is based upon work supported by, or in part by, the US Army Research Laboratory and the US Army Research Office.

[2] A. Cazenave and R. S. Nerem, "Present-day sea level change: Observations and causes," *Reviews of Geophysics* 42 (2004), RG3001, 20 PP.

[3] R. Schubert et al. , *The future oceans—warming up, rising high, turning sour* (Berlin: German Advisory Council on Global Change, 2006).

[4] The Global Climate Observing System is a joint undertaking of the World Meteorological Organization, the Intergovernmental Oceanographic Commission of the United Nations Educational Scientific and Cultural Organization, the United Nations Environment Programme, and the International Council for Science. As a system of climate-relevant observing systems, it constitutes, in aggregate, the climate observing component of the Global Earth Observation System of Systems.

[5] Global Climate Observing System, *Implementation Plan for the Global Observing System for Climate in Support for the UNFCCC* [United National Framework Convention on Climate Change] (World Meteorological Organization, 2010).

[6] Committee on National Security Implications of Climate Change for US Naval Forces, *National Security Implications of Climate Change for US Naval Forces* (Washington, D.C.: The National Academies Press, 2011).

[7] R.M. Duren and C.E. Miller, "Towards Robust Global Greenhouse Gas Monitoring," *Greenhouse Gas Measurement and Management* (2011).

[8] The US Group on Earth Observations is chartered by the Office of Science and Technology Policy to prioritize and offer guidance for elements of a US global observing system as part of our nation's contribution to the international Global Earth Observational System of Systems. USGEO plays a role in identifying critical gaps in US observational capability and makes recommendations for closing those gaps.

[9] The US Group on Earth Observations, *Achieving and Sustaining Earth Observations: A Preliminary Plan Based on a Strategic Assessment* (2010).

[10] The former National Polar-orbiting Operational Environmental Satellite System program is moving forward as the separate Joint Polar Satellite System and the Defense Weather Satellite System programs which will address some, but not all, of the ECVs. Most ECVs will continue to be delivered by exploratory science assets for the foreseeable future.

[11] America's Climate Choices: Panel on Advancing the Science of Climate Change, National Research Council, *Advancing the Science of Climate Change* (Washington, D.C.: The National Academies Press, 2010).

[12] http://cmip-pcmdi.llnl.gov/cmip5/

[13] S. Solomon, D. Qin, M. Manning, Z. Chen, M. Marquis, K.B. Averyt, M.Tignor and H.L. Miller, eds., "Summary for Policymakers," in *Climate Change 2007: The Physical Science Basis*, Contribution of Working Group I to the Fourth Assessment Report of the Intergovernmental Panel on Climate Change (Cambridge University Press: Cambridge, United Kingdom and New York, NY, 2007).

[14] Filippo Giorgi, Colin Jones and Ghassem R. Asrar, "Addressing climate information needs at the regional level: the CORDEX framework," *World Meteorological Organization Bulletin* 58:3 (July 2009).

[15] S. Solomon, D. Qin, M. Manning, Z. Chen, M. Marquis, K.B. Averyt, M. Tignor and H.L. Miller, eds., *Climate Change 2007: The Physical Science Basis*, Contribution of Working Group I to the Fourth Assessment Report of the Intergovernmental Panel on Climate Change (Cambridge University Press: Cambridge, United Kingdom and New York, NY, 2007).

[16] See: D.S. Arndt, M. O. Baringer, and M. R. Johnson, eds., "State of the Climate in 2009," *American Meteorological Society Bulletin* 91, no. 7 (2010): S1-S224; H. M. Füssel, "The risks of climate change: A synthesis of new scientific knowledge since the finalization of the IPCC Fourth Assessment Report (AR4)," Background note to the World Development Report 2010 (Germany: Potsdam Institute for Climate Impact Research, 2010); C. Richardson (Chair), *Synthesis Report from Climate Change, Global Risks, Challenges and Decisions* (Copenhagen: University of Copenhagen, 2009); M. Rummukainen, et al., *Physical Climate Science since IPCCAR4: A brief update on new findings between 2007 and April 2010* (Denmark: Danish Meteorological Institute, 2010); R. van Dorland, et al.,

News in Climate Science Since IPCC 2007: Topics of interest in the scientific basis of climate change (Royal Netherlands Meteorological Institute (KNMI) and others, 2009); Netherlands Environmental Assessment Agency, *News in Climate Science and Exploring Boundaries: A Policy brief on developments since the IPCCAR4 report in 2007*, PBL 500114013 (The Netherlands: Netherlands Environmental Assessment Agency, 2009).

[17] Solomon et al., eds., (2007).

[18] Rummukainen et al. (2010).

[19] J. Hanson, R. Ruedy, M. Sato, and K. Lo, "Global Surface Temperature Change," *Reviews of Geophysics* 48, RG4004 (2010) 29 PP. Copyright 2010 American Geophysical Union. Reproduced by permission of American Geophysical Union.

[20] Solomon et al., eds. (2007) and Netherlands Environmental Assessment Agency (2009).

[21] Solomon et al., eds. (2007).

[22] S. Hastenrath, D. Polzin, and C. Mutai, "Diagnosing the Droughts and Floods in Equatorial East Africa during Boreal Autumn 2005-08," *Journal of Climate* 23 (2010): 813-817, DOI: 10.1175/2009JCLI3094.1.

[23] A. Dai, "Drought under global warming: a review," *Climate Change* 2, no. 1(2011). 45-65, Doi: 10.1002/wcc.81.

[24] van Dorland et al. (2009).

[25] B.R. Mackenzie and D .Schiedek, "Daily ocean monitoring since the 1860's shows record warming of northern European seas," *Global Change Biology* 13, no. 7 (2007): 1335-1347.

[26] A.P. Williams and C. A. Funk, "A westward extension of the warm pool leads to a westward extension of the Walker circulation, drying eastern Africa," *Climate Dynamics* (2011), Doi:10.1007/s00382-010-0984-y.

[27] Sep 2010 minus Aug 2010 = -0.028°C.

[28] S. Rahmstorf, "A new view on sea level rise," *Nature Reports Climate Change* 4 (2010): 44-45.

[29] See: J.L. Chen, C. R. Wilson, and B. D. Tapley, "Satellite gravity measurements confirm accelerated melting of Greenland ice sheet," *Science* 313 (2006): 1958-1960; D.C. Slobbe, R.C. Lindenbergh and P. Ditmar, "Estimation of volume change rates of Greenland's ice sheet form ICEsat data using overlapping footprints," *Remote Sensing of Environment* 112 (2008): 4204-4213; I. Velicogna, "Increasing rates of ice mass loss from the Greenland and Antarctic ice sheets revealed by GRACE," *Geophysical Research Letters* 36 (2009): L19503, Doi:10.102912009GL040222; and van Dorland et al. (2009).

[30] T.L. Mote, "Greenland surface melt trends 1973-2007: Evidence of a large increase in 2007," *Geophysical Research Letters* 34, no. 22 (2007).

[31] B. Wouters et al., "GRACE observes small-scale mass loss in Greenland," *Geophysical Research Letters* 35 (2008), L20501, Doi:10.1029/2008GL034816.

[32] E.J. Steig , D.P. Schneider, S.D. Rutherford, M.E. Mann, J.C. Comiso, and D.T. Schindell, "Warming of the Antarctic ice-sheet surface since the 1957 International Geophysical Year," *Nature* 457, no. 7228 (2009): 459-462, Doi:10.1038/nature07669, PMID 19158794.

[33] E. Rignot, J.L. Bamber, M. R. van den Broeke, C. Davis, Y. Li, W. J. van de Berg, and E. van Meijgaard, "Recent Antarctic ice mass loss from radar interferometry and regional climate modeling," *Nature Geoscience* 1, no. 2 (2007): 106-110; and van Dorland et al. (2009).

[34] Rignot et al. (2008).

[35] Rignot et al. (2008).

[36] United Nations Environment Programme, World Glacier Monitoring Service, *Global Glacier Changes: Facts and Figures* (2008).

[37] F. M. Meier, et al., "Glaciers dominate eustatic sea-level rise in the 21st century," *Science* 317 (2007): 1064-1067.

[38] See: J. Boé, et al., "September sea ice cover in the Arctic Ocean projected to vanish by 2100,"*Nature Geoscience* 2 (2009): 341-343, Doi:10.1038/ngeo467; and M. Wang and J.E. Overland, "A sea ice free summer Arctic within 30 years," *Geophysical Research Letters* 36 (2009), L07502.

[39] Netherlands Environmental Assessment Agency (2009).

[40] J.C. Comiso et al., "Accelerated decline in the Arctic sea ice cover," *Geophysical Research Letters* 35 (2008), L01703, Doi:10.1029/2007GL031972.

[41] J. Stroeve, M. M. Holland, W. Meier, T. Scambos, and M. Serreze, "Arctic sea ice decline: Faster than forecast" *Geophysical Research Letters* 34 (2007): 9501.

[42] Netherlands Environmental Assessment Agency (2009).

[43] K.E. Thornton, "An update on human-induced climate change," Testimony of Dr. Kevin E. Trenberth before the US Senate Committee on Environment and Public Works, July 22, 2008

[44] Thornton (2008).

[45] K.M. Walter, S. A. Zimov, J. P. Chanton, D. Verbyla, and F. S. Chapin III, "Methane bubbling from Siberian thaw lakes as a positive feedback to climate warming," *Nature* 443 (2006): 71-75.

[46] V.E. Romanovsky, S.L. Smith, S. L. and H.H. Christiansen, "Permafrost thermal state in the polar Northern Hemisphere during the international polar year 2007-2009: a synthesis," *Permafrost and Periglacial Processes* 21 (2010): 106-116, Doi: 10.1002/ppp.689.

[47] "Warming World: Impacts by degree," based on the National Research Council report, *Climate Stabilization Targets: Emissions, Concentrations, and Impacts over Decades to Millennia* (Washington, D.C.: The National Academies Press, 2010).

[48] M. Rigby et al., "Renewed growth of atmospheric methane," *Geophysical Research Letters* 35 (2008), L22805, Doi:10.1029/2008GL036037.

[49] E. A. G. Schuur, J. Bockheim, J. G. Canadell, E. Euskirchen, C. B. Field, S. V. Goryachkin, S. Hagemann, P. Kuhry, P. M. Lafleur, H. Lee, G. Mazhitova, F. E. Nelson, A. Rinke, V. E. Romanovsky, N. Shiklomanov, C. Tarnocai, S. Venevsky, J. G. Vogel, and S. A. Zimov, "Vulnerability of permafrost carbon to climate change: Implications for the global carbon cycle," *BioScience* 58, no. 8 (2008): 701-714, Doi:10.1641/BS80807.

[50] Rummukainen et al. (2010).

[51] van Dorland et al. (2009) citing Dorrepaal et al. (2009), full reference not available.

[52] A. Dai (2011).

[53] E.J. Burke, S.J. Brown, and N. Christidis, "Modeling the Recent Evolution of Global Drought and Projections for the Twenty-First Century with the Hadley Centre Climate Model," *Journal of Hydrometeorology* 7 (2006): 1113-1125.

[54] A. Dai, K.E. Trenberth, and T. Qian, "A global data set of Palmer Drought Severity Index for 1870-2002: Relationship with soil moisture and effects of surface warming" *Journal of Hydrometeorology* 5 (2004): 1117-1130.

[55] J. Sheffield, *Global drought in the 20th and 21st centuries: Analysis of retrospective simulations and future projections of soil moisture*, PhD thesis (Princeton, New Jersey: Princeton University, 2008).

[56] Burke et al. (2006).

[57] The IPCC Special Report on Emissions Scenarios (SRES) scenarios are projections of future greenhouse gas emissions used in climate models. The scenarios are based on storylines, each of which makes different assumptions about future demographic, economic, and technological driving forces of greenhouse gas and sulphur emissions.

[58] Arndt et al. (2009).

[59] Trenberth (2008).

[60] Arndt et al. (2010).

[61] Determined using the GHCN-Monthly dataset for the following 3-month seasons: (a) Dec 2008 to Feb 2009, (b) Mar to May 2009, (c) Jun to Aug 2009, and (d) Sep to Nov 2009. Seasonal anomalies were determined relative to the 1961-90 means, with at least two-thirds (66%) of the years without missing data required during the base period. D. S. Arndt, M. 0. Baringer, and M. R. Johnson, eds., "State of the Climate in 2009," *Bulletin of the American Meteorological Society* 91, no. 7 (2010): S1-S224. (c)American Meteorological Society. Reprinted with permission.

[62] V. Alavian, H.M. Qaddumi, Eric Dickson, S.M. Diez, A. V. Danilenko, R.F. Hirji, G. Puz, C. Pizarro, M. Jacobsen, and B. Blankespoor, *Water and Climate Change: Understanding the Risks and Making Climate-smart Investment Decisions* (The World Bank, November 2009).

[63] Alavian et al. (2009).

[64] Taking the flow that is exceeded 10% of the time (q10)· which means there is a 90% chance in each time period of a flow lower than this. A decrease in q10 means that the likelihood of high flows and floods will increase. The relative changes from historical values provide an "indicator" of the projected change in floods.

[65] F. Gutiérrez et al.· "Recent advances in landslide investigation: Issues and perspectives·" Guest editorial, *Geomorphology* 124 (2010): 95-101.

[66] S.J. Brown, J. Caesar, and C. A. T. Ferro, "Global changes in extreme daily temperature since 1950," *Journal of Geophysical Research-Atmospheres* 113 (2008).

[67] A Sterl et al., "When can we expect extremely high surface temperatures?" *Geophysics Research Letters* 35 (2008), L14703, Doi:10.1029/2008GL034071.

[68] See: K. Emanuel, "Increasing destructiveness of tropical cyclones over the past 30 years,"*Nature* 436 (2005a): 686-688; D. W. Landsea, B. A. Harper, K. Hoarau and J.A. Knaff, "Can we detect trends in extreme tropical cyclones?" *Science* 313 (2006): 452-454; K. E. Trenberth and D.J. Shea, "Atlantic hurricanes and natural variability in 2005," *Geophysical Research Letters* 33 (2006), L12704 10.1029/2006GL026256; and P.J. Webster, G. J. Holland, J.A. Curry, and H.-R. Chang, "Changes in tropical cyclone number, duration, and intensity in a warming environment," *Science* 309 (2005): 1844-1846.

[69] (A) The numbers of North Atlantic tropical cyclones for each maximum wind speed shown on the horizontal axis. (B) The proportional increase by cyclone (hurricane) category (1 - least intense; 5 - most intense) arising from increases in maximum wind speeds of 1, 3 and 5 m/s.

[70] The 5-year running mean normalized Atlantic main development region SST indices from three different reconstructions of SST overlaid on Atlantic tropical cyclone counts. Blue curves are the three main development region SST reconstructions. The heavy red curve is the base-case tropical cyclone count. The light red curve is the unadjusted tropical cyclone count. The orange curves are the US land falling tropical storms and hurricane series counts from the North Atlantic hurricane database.

[71] G.A. Vecchi and T.R. Knutson, "On Estimates of Historical North Atlantic Tropical Cyclone Activity," *Journal of Climate* 21 (2008): 3580-3600.

[72] See: Rummukainen et al. (2010); U. Löptien et al., "Cyclone life cycle characteristics over the Northern Hemisphere in coupled GCMs," *Climate Dynamics* 31 (2008): 507- 532, Doi:10.1007/s00382-007-0355-5; and L. Bengtsson et al., "Will extratropical storms intensify in a warmer climate?" *Journal of Climate* 22 (2009): 2276-2301, Doi:10.1175/ 2008JCLI2678.1.

[73] See: J.H. Yin, "A consistent poleward shift of the storm tracks in simulations of 21st century climate," *Geophysical Research Letters* 32 (2005), L18701, Doi:10.1029/2005GL023684; and G. Gastienau and B.J. Soden, "Model projected changes of extreme wind events in response to global warming," *Geophysical Research Letters* 36 (2009), L10810, Doi:10.1029/2009GL037500.

[74] See: E.P. Salathé Jr, "Influences of a shift in North Pacific storm tracks on western North American precipitation under global warming," *Geophysical Research Letters* 33 (2006), L19820, Doi:10.1029/2006GL026882; and J. Jiang, and W. Perrie, "Climate change effects on North Atlantic cyclones,"*Journal of Geophysical Research Letters* 113 (2008), D09102, Doi:10.1029/2007JD008749.

[75] Riley Duran. Presentation to the DSB Task Force, July 2010. Jet Propulsion Laboratory.

[76] See: Solomon et al., eds. (2007) and Trenberth (2008).

[77] A. Cazenave, K. Dominh., S. Guinehut, E. Berthier, W. Llovel, G. Ramilien, M. Ablain, and G. Larnicol, "Sea level budget over 2003-2008. A reevaluation from GRACE space gravimeter, satellite altimetry and Argo," *Global and Planetary Change* 65 (2009): 83-88.

[78] S. Rahmstorf, "A new view on sea level rise," *Nature Reports Climate Change* 4 (2010): 44-45.

[79] J.T. Overpeck and J.L.Weiss, "Projections of future sea level becoming more dire," *Proceedings of the National Academy of Sciences* 106 (2009): 21461-21462, Doi: 101073/pnas.0912878107

[80] Meehl et al., "How much more warming and sea level rise?" *Science* 307 (2005) 1769-1772.

[81] N. Mabey, N., J. Gulledge, B. Finel, and K. Silverthorne. *Degrees of Risk: Defining a Risk Management Framework for Climate Security* (London, UK: Third Generation Environmentalism Ltd, February 2011).

[82] See: Solomon et al. (2009); D. Archer and V. Brovkin, "The millennial atmospheric lifetime of anthropogenic CO_2," *Climatic Change* 90, no. 3 (2008): 283-297; H.D. Matthews and K. Caldeira, "Stabilizing climate requires near zero emissions," *Geophysical Research Letters* 35 (2008), L04705; and M. Eby, K. Zickfeld, A. Montenegro, D. Archer, K. J. Meissner, and A. J. Weaver, "Lifetime of anthropogenic climate change: millennial time scales of potential CO_2 and surface temperature perturbations," *Journal of Climate* 22 (2009): 2501-2511.

[83] See: G.K. Plattner, "Long-term commitment of CO_2 emissions on the global carbon cycle and climate," IOP Conference Series: Earth and Environmental Sciences 6 (2009). 042008. available online at: http://www.iop.org/EJ/toc/1755-1315/6/4; and S. Solomon, G.K. Plattner, R. Knutti, and P. Friedlingstein, "Irreversible climate change due to carbon dioxide emissions," *Proceedings of the National Academy of Sciences* 106 (2009): 1704-1709.

[84] T.M. Lenton and H.J. Schellnhuber, "Tipping the scales," *Nature Reports Climate Change* 1 (2007): 97-98.

[85] There is one more tipping element not shown, the Indian Summer Monsoon, because its critical threshold cannot be meaningfully related to global warming. The temperature ranges given here are from reviewing studies in the literature and conversations with individual experts.

[86] A.R. Ganguly, K. Steinhaeuser, E.S. Parish, S.C. Kao, A.W. King, M. Branstetter, A. Sorokine, *Climate Change Support for the United States Department of Defense (US DOD),* 2010 Quadrennial Defense Review report (Oak Ridge National Laboratory Climate Change Science Institute, 2010). The research was funded by the US Office of the Secretary of Defense. The Oak Ridge National Laboratory is managed by UT-Battelle, LLC, for the US Department of Energy.

[87] M.E. Shongwe, G.J. van Oldenborgh, B.J.J.M. van den Hurk, B. de Boer, C.A.S. Coelho, and M.A. van Aalst, Projected changes in extreme precipitation in Africa under global warming. Poster, Workshop on metrics and methodologies of estimation of extreme climate events, 27-29 September 2010, UNESCO headquarters, Paris, France.

[88] M.E. Shongwe, G.J. van Oldenborgh, B.J.J.M. van den Hurk, and M. van Aalst, "Projected changes in mean and extreme precipitation in Africa under global warming, Part II: East Africa," Submitted *Journal of Climate* (2010).

[89] International Bank for Reconstruction and Development, *Water and Climate Change: Understanding the Risks and Making Climate-Smart Investment Decisions* (The World Bank: Publication, November 2009).

[90] Shongwe et al. (2010).

[91] P. Meier, Doug Bond, Joe Bond, "Environmental influences on pastoral conflict in the Horn of Africa,"*Political Geography* 26, no. 6 (2007): 716-735.

[92] A. Agoumi, *Vulnerability of North Africa Counties to Climate Change: Adaptation and Implementation Strategies for Climate Change* (International Institute for Sustainable Development (2003).

[93] R. Kasei and B. Diekkriigger, "Drought frequency in the Volta Basin of West Africa," *Sustainability Science* 5 (2010): 89-97, Doi 10.1007/s11625-009-0101-5.

[94] US Army Corps of Engineers, *The Impact of Water as a United States Government Security Consideration for the US Africa Command* (September 29, 2010). Report prepared for the US Africa Command.

[95] A. Dai (2011).

[96] RANET, Radio and Internet for the communication of hydro-meterollogical information for rural development, "is a collaborative effort of many national hydro- meteorological services, nongovernmental organizations, and communities. These varied partners come together to make weather, water, and climate information available to rural and remote populations, which are often most in need of environmental forecasts, observations, and warnings." www.ranetproject.net [Accessed July 22, 2011].

[97] Adapted from B. L. Turner II, Roger E. Kaspersonb, Pamela A. Matsone, James J. McCarthy, Robert W. Corell, Lindsey Christensene, Noelle Eckley, Jeanne X. Kaspersonb, Amy Luerse, Marybeth L. Martellog, Colin Polsky, Alexander Pulsipher, and Andrew Schiller, "A framework for vulnerability analysis in sustainability science," *Proceedings of the National Academy of Sciences* 100, no. 14 (July 8, 2003) 8074-8079.

[98] International Bank for Reconstruction and Development, *Convenient Solutions to an Inconvenient Truth: Ecosystem-based Approaches to Climate Change* (Washington, D.C.: World Bank, 2009).

[99] US Army Corps of Engineers (2010).

[100] *UN Water, A Snapshot of Drinking Water and Sanitation in Africa* (2008). A regional perspective based on new data from the World Health Organization/UNICEF Joint Monitoring Programme for Water Supply and Sanitation. Prepared for AMCOM as a contribution to the 11th Summit of Heads of State and Government of the African Union With special theme: Meeting the Millennium Development Goal on Water and Sanitation 30 June to 1 July, 2008.

[101] World Health Organization, *Ten facts about water security* (World Health Organization, 2009).

[102] J. Delli Prescoli, "Water Security, Global Water Issues and Climate Change Water: Key to Regional Stability," Keynote speech at International Water Symposium, Korean Environment Institute, Climate Change Water Security, July 20, 2010, Korea Chamber of Commerce and Industry Seoul Korea.

[103] UNEP/GRID-Arendal, Water availability in Africa, UNEP/GRID-Arendal Maps and Graphics Library, http://maps.grida.no/go/graphic/water_availability_in_africa (Accessed 29 July 2011). Cartographer: Digout, Delphine, based on a sketch by Philippe Rekacewicz; UNEP/GRID-Arendal. Sources: United Nations Economic Commission for Africa (UNECA), Addis Ababa; Global Environment Outlook 2000 (GEO 2000), UNEP, Earthscan, London, 1999.

[104] Suhas P. Wani, J. Rockstrom, and T. Oweis, eds, *Rainfed Agriculture: Unlocking the Potential* (United Kingdom: International Water Management Institute, 2009).

[105] K.D. White, Army Corps of Engineers. Presentation to the Task Force on December 17, 2010. Based on M. van Aalst, M Hellmuth, and D Ponzi, "Come Rain or Shine: Integrating Climate Risk Management into African Development Bank Operations," Working Paper No 89 (Tunis: African Development Bank, 2007). Source: World Bank.

[106] *Africa Environment Outlook 2: Our environment, our wealth* (United Nations Environment Programme, 2006).

[107] The Water for the Poor Act was enacted in 2005 by the Bureau of Oceans and International Environmental and Scientific Affairs within the Department of State. The Act authorizes the use of foreign aid to expand the role of global water and sanitation programs in US foreign policy. Additionally, it allows the secretary of state to create international programs that promote increased levels of clean water and sanitation around the world.

[108] In Section 1207 of the FY 2006 National Defense Authorization Act, Congress authorized DoD to transfer up to $100 million per year to the Secretary of State for "reconstruction, security, or stabilization assistance to a foreign country." Congress then reauthorized this authority in Section 1210 of the FY 2008 Defense Authorization Act. These funds have hence been informally referred to as "1207" or "1210" funds.

[109] Peter Ashton, Aquatic Ecosystems Research Group, Natural Resources and the Environment, CSIR, Pretoria, South Africa. Presentation on Key Challenges Facing Water Governance in

Africa, presented at: South African Institute of International Affairs Symposium: "The Second Scramble for Africa: Lifting the Resources Curse": SAIIA, Johannesburg, 29 November 2007. Data from: Food and Agriculture Organization of the United Nations (2005). Aquastat Database.

[110] J. Delli Priscoli and A.T. Wolf, *Managing and Transforming Water Conflicts* (Cambridge University Press, 2008).

[111] DOD/Combatant Commander Water Security Program Strategy, April 2010. Available from Erik Fleischner, HQUSACE LNO, United States European Command.

[112] Mabey et al. (2011).

[113] US Army Corps of Engineers (2010).

[114] DOD/Combatant Commander Water Security Program Strategy, April 2010.

[115] East Africa: Uganda, DR Congo Head Off Dispute as River Alters Border. allAfrica.com, November 9, 2009.

[116] Presentation by Dr. K. White to the DSB Task Force, citing (Freeman, JAWRA 2000).

[117] White Paper on Defence, 1996.

[118] For a discussion of the nature of puzzles and mysteries see: Malcom Gladwell, "Open Secrets" in *What the Dog Saw and Other Adventures* (New York, NY: Little, Brown and Company, 2009).

[119] National Intelligence Council, *North Africa: Impact of Climate Change to 2030: Geopolitical Implications* (December 2009) p. 11.

[120] G. Fischer, M. Shah, F.N. Tubiello, and H., van Velhuizen, "Socio-economic and climate change impacts on agriculture: an integrated assessment, 1990-2080," *Philosophical Transactions of the Royal Society B*, 360 (1463): 2067-2083, Doi: 10.1098/rstb.2005.1744, November 2005.

[121] P.L. Thornton, P.G. Jones, T. Owiyo, R.I. Kruska, M. Herrer, P. Kristjanson, A. Notenbaert, N. Bekele and A. Omolo, with contributions from V. Orindi, B. Otiende, A. Ochieng, S. Bhadwal, K. Anantram, S. Nair, V. Kumar, and U. Kulkar, *Mapping climate vulnerability and poverty in Africa* (2006). Report to the Department for International Development.

[122] Cheung et al., "Large-scale redistribution of maximum fisheries catch potential in the global ocean under climate change," *Global Change Biology* 16 (2010): 24-35.

[123] C. Nellemann, M. MacDevette, T. Manders, B. Eickhout, B. Svihus, A. G. Prins, and B. P. Kaltenborn eds., *The Environmental Food Crisis: The Environment's Role in Averting Future Food Crises* (UNEP, February 2009), http://www.grida.no/publications/rr/food-crisis/

[124] K. D. White, US Army Corps of Engineers. Presentation to the DSB Task Force, December /7, 20/0.

[125] Susmita Dasgupta, Benoit Laplante, Siobhan Murray, and David Wheeler, *Sea Level Rise and Storm Surges: A Comparative Analysis of Impacts in Developing Countries*, Public Research Working Paper 490/ (World Bank, April 2009).

[126] UN-HABITAT, *African Cities at Risk due to Sea-Level Rise* (Global Urban Observatory, 2008).

[127] Henri E.Z. Tonnang, Richard Y.M. Kangalawe, and Pius Z. Yanda, "Predicting and mapping malaria under climate change scenarios: the potential redistribution of malaria vectors in Africa," *Malaria Journal* 9, no. 111 (2010), Doi: 10.1186/1475-2875- 9-111. Published online April 23, 2010.

[128] S. Hales et al., "Potential effect of population and climate changes on global distribution of dengue fever: an empirical model," *The Lancet* 360 (2002): 830-834.

[129] Joshua W. Busby, Todd G. Smith, Kaiba L. White, and Shawn M. Strange, *Locating Climate Insecurity: Where are the Most Vulnerable Places in Africa*? (Austin, Texas: Robert S. Strauss Center for International Security and Law, 2010). This material is based upon work supported by, or in part by, the US Army Research Laboratory and the US Army Research Office under contract/grant number W911NF-09-1-0077.

[130] P. Watkiss, T. Downing, and J. Dys2ynski, *AdaptCost Project: Analysis of the Economic Costs of Climate Change Adaptation in Africa* (Nairobi: UNEP, 2010).

[131] Attempts at geoengineering, on a global scale, to modify or reverse climate change are especially dangerous. The knowledge of the complete range of factors affecting Earth's climate is so rudimentary that climate engineering is virtually guaranteed to result in worse problems than any Mother Nature ever dreamed up.

[132] James Rodger Fleming, *Fixing the Sky: The Checkered History of Weather and Climate Control* (New York, NY: Columbia University Press, September 2010).

[133] Swedish Ministry of Defense, *On Connecting Climate Change with Security and Armed Conflict* (Department of Defence Analysis, September 2010).

[134] Department of State, http:11www.state.gov1g1oes1hub1 [Accessed 8 August 2011].

[135] Sherri Wasserman Goodman, Deputy Under Secretary of Defense for Environmental Security, Statement Before the Subcommittee on Installation and Facilities, May 13, 1993.

[136] Lori Dengler, *Strategic Implementation Plan for Tsunami Mitigation Projects*, NOAA Technical Memorandum ERL PMEL-113 (National Oceanic and Atmospheric Administration, 1998) www.pmel.noaa.gov/pubs/PDF/deng2030/deng2030.pdf [Accessed August 2011].

[137] Examples of these studies include: *America's Climate Choices* (2011), *National Security Implications of Climate Change for US Naval Forces* (2011), *Advancing the Science of Climate Change* (2010), *Limiting the Magnitude of Future Climate Change* (2010), *Adapting to the Impacts of Climate Change* (2010).

[138] White House, *National Security Strategy*, May 2010, p. 47.

[139] President Barack Obama's Nobel Peace Prize Acceptance Speech, 2009.

[140] Gladwell (2009).

[141] America's Climate Choices: Panel on Informing Effective Decisions and Actions Related to Climate Change, National Research Council, *Informing an Effective Response to Climate Change* (Washington, D.C.: The National Academies Press, 2010).

[142] Mabey et al. (2011).

[143] At the time of this writing (March 25, 2011) operators at the Fukushima nuclear plant in Japan are still struggling to gain control of cooling systems of reactors and fuel storage pools damaged by a 46 foot tsunami triggered by an offshore magnitude 9.0 earthquake two weeks ago. According to press reports, the Fukushima plant's tsunami protection features (modest augmentation around a 13 foot natural seawall) were designed using a "deterministic" rather than probabilistic approach (i.e., based on recent historical experience with tsunamis rather than addressing reasonable worst- case scenario which transpired). The risk of underestimating events with so called "long tail" probabilistic distributions is exacerbated for climate change scenarios.

[144] Lenton et al., "Tipping elements in the earth's climate system," *Proceedings of the National Academy of Sciences* 105, no. 6 (2008): 1786-1793.

[145] Tsunami threat potential, which varies strongly with local tides, winds, topography, and bathymetry can arguably be amplified by rising sea levels.

[146] See: Hoerling et al. (2004); S. Janicot, S. Trzaska, and I. Poccard, "Summer Sahel-ENSO teleconnection and decadal time scale SST variations," *Climate Dynamics* 18 (2001): 303-320; and Raicich et al. "Teleconnections Between Indian Monsoon And Sahelm Rainfall And The Mediterranean," *International Journal of Climatology* 23 (2003): 173-186.

[147] Davis and Caldiera, "Consumption-based accounting of CO_2," *Proceedings of the National Academy of Sciences* (2010).

[148] R.J.T. Klein, S. Huq, F. Denton, T.E. Downing, R.G. Richels, J.B. Robinson, and F.L. Toth, "Inter-relationships between adaptation and mitigation" in *Climate Change 2007: Impacts, Adaptation and Vulnerability*. Contribution of Working Group II to the Fourth Assessment Report of the Intergovernmental Panel on Climate Change, M.L. Parry, O.F. Canziani, J.P. Palutikof, P.J. van der Linden and C.E. Hanson, Eds. (Cambridge, UK: Cambridge University Press, 2007) pp. 745-777.

[149] Miles et al., "An approach to designing a national climate service," *Proceedings of the National Academy of Sciences* (2006).

[150] http://www.globalchange.gov/; http://www.wmo.int/pages/themes/climate; http://www.earth observations.org

[151] Morello, "Agency Proposes Climate Service to Spur Adaptation," *New York Times*, Feb 9, 2010, http://www.nytimes.com/cwire/2010/02/09/09climatewire-agencywill-create-national-climate-service-63603.html

[152] Advertised products from the NOAA Climate Service include (for the US): Inundation maps for coastal communities that reflect the best available information on sea level rise and changing patterns of coastal storms; heat projections to help managers plan future energy and health services needs; climate and precipitation models to help farmers know the impact of a changing climate on their crops; relevant historical climate data and data from state-of-the-art climate models to inform investment and planning for businesses and local governments; routine vulnerability and risk assessments for climate-sensitive regions and sectors.

[153] http://www.noaa.gov/climate.html

[154] http://www.doi.gov/whatwedo/climate/strategy/CSC-Map.cfm

[155] J. M. Murphy, D. M. H. Sexton, D. N. Barnett, G. S. Jones, M. J. Webb, and M. Collins, "Quantification of modelling uncertainties in a large ensemble of climate change simulations," *Nature* 430, no. 7001 (2004): 768-772.

[156] P. Gleckler, R. Ferraro, and D. E. Waliser, Better use of satellite data in evaluating climate models contributing to CMIP and assessed by IPCC, Joint DOE-NASA workshop, Lawrence Livermore National Laboratory, October 12-13, 2010, EOS, In Press.

[157] A value of -0.2 means that the model has an error 20% smaller than the typical model error for that quantity. Each grid square is split by a diagonal in order to show the relative error with respect to both the primary (upper left triangle) and the alternate (lower right triangle) reference data sets. Variables included: hfls and hfss - latent and sensible heat flux; rxxx - various TOA and surface quantities of long wave and shortwave radiation including cloud radiative forcing; ts, clt - surface temperature, total cloud cover; pr, prw and psl are precipitation, precipitable water and sea level pressure; remaining variables are surface stress and winds and winds, humidity, temperature and heights at a number of pressure levels.

[158] T. Reichler and J. Kim, "How well do coupled models simulate today's climate?" *Bulletin of the American Meteorological Society* 89, no. 3 (2008): 303.

[159] Best performing models have low I2 values and are located toward the left. Circle sizes indicate the length of the 95% confidence intervals. Letters and numbers identify individual models (not shown); flux-corrected models are labeled in red. Grey circles show the average I2 of all models within one model group. Black circles indicate the I2 of the multi-model mean taken over one model group. The green circle (REA) corresponds to the I2 of the National Centers for Environmental Prediction/National Center for Atmospheric Research reanalyses. (right) Climate variables and validation data.

[160] See: D. Waliser, K. W. Seo, S. Schubert, and E. Njoku, "Global water cycle agreement in the climate models assessed in the IPCC AR4," *Geophysical Research Letters* 34, no. 16 (2007); and D. Waliser et al., "Cloud ice: A climate model challenge with signs and expectations of progress,"*Journal of Geophysical Research-Atmospheres* 114 (2009).

[161] K.E. Trenberth, A. G. Dai, R. M. Rasmussen, and D. B. Parsons, "The changing character of precipitation," *Bulletin of the American Meteorological Society* 84, no. 9 (2003): 1205.

[162] NASA's Earth Observing System Data and Information System (http://nasadaacs.eos.nasa.gov/)

[163] USGS's Earth Resources Observation and Science Center (http://eros.usgs.gov/)

[164] IPCC's Data Distribution Center (http://www.ipcc-data.org/)

[165] Program for Climate Model Diagnosis and Intercomparison (http://wwwpcmdi.llnl.gov/)

[166] National Integrated Drought Information System (http://www.drought.gov)

[167] Famine Early Warning System Network (http://www.fews.net)
[168] US Department of Agriculture Foreign Agriculture Service's Crop Explorer (http://www.pecad.fas.usda.gov/cropexplorer/)
[169] Regional Visualization and Monitoring System (http://www.servir.net/en/)
[170] Busby et al. (2010).
[171] Anthony Nyong, *Climate change & Africa's security: the role of the African Development Bank,* presentation to the Defense Science Board Task Force on Climate Change, November 8, 2010.
[172] United States Global Change Research Program, *Our changing planet*, Table 1. FY2007-2009 Climate change Science Program Budget by Goal and Focus Area (2010).
[173] Mabey et al. (2011).
[174] The White House Council on Environmental Quality, *Progress Report of the Interagency Climate Change Adaptation Task Force: Recommended Actions in Support of a National Climate Change Adaptation Strategy* (Washington, D.C.: Executive Office of the President, 2010).
[175] The White House Council on Environmental Quality (2010).
[176] http://www.globalchange.gov/what-we-do/assessment/nca-overview
[177] "U.S. National Climate Assessment Objectives, Proposed Topics, and Next Steps," *Federal Register* 60, no. 210 (September 7, 2010).
[178] George Backus, Thomas Lowry, Drake Warren, Mark Ehlen, Geoffrey Klise, Verne Loose, Len Malczynski, Rhonda Reinert, Kevin Stamber, Vince Tidwell, Vanessa Vargas, and Aldo Zagonel, *Assessing the Near-Term Risk of Climate Uncertainty: Interdependencies among the US states*, SAND2010-2052 (Albuquerque, NM: Sandia National Laboratory, May 2010).
[179] Meier et al., *Climate Change and Conflict* (2007) pp. 716-735.
[180] America's Climate Choices: *Advancing the Science of Climate Change* (2010).
[181] The White House Council on Environmental Quality (2010).
[182] George Backus et al. (20/0).
[183] The White House Council on Environmental Quality (2010).
[184] United States Global Change Research Program, *Our Changing Planet: The US Global Change Research Program for Fiscal Year 2011.*
[185] The White House Council on Environmental Quality (2010).
[186] Lenton et al. (2008).
[187] America's Climate Choices: *Advancing the Science of Climate Change* (2010).
[188] Lenton and Schellnhuber (2007).
[189] Alley et al., "Abrupt increase in Greenland snow accumulation at the end of the Younger Dryas event," *Nature* 362 (1993).
[190] America's Climate Choices: *Advancing the Science of Climate Change* (2010).
[191] Modified from Figure 2.4 on p. 35 of: Nick Mabey, Jay Gulledge, Bernard Finel, and Katherine Silverthorne, "Degrees of Risk: Defining a Risk Management Framework for Climate Security," February 2011.
[192] Susan Solomon, Gian-Kasper Plattner, Reto Knutti, and Pierre Friedlingstein, Physical Sciences: "Environmental Sciences: Irreversible climate change due to carbon dioxide emissions," *Proceedings of the National Academy of Sciences* 106, no. 6 (2009): 1704-1709. Published ahead of print January 28, 2009.
[193] *Geoengineering the climate: Science, governanace and uncertainity* (Royal Society, 2009).
[194] America's Climate Choices: *Advancing the Science of Climate Change* (2010)
[195] Government Accountability Office, *Climate Change: A Coordinated Strategy Could Focus Federal Geoengineering Research and Inform Governance Efforts,* GAO-10-903 (September 2010).
[196] Government Accountability Office (2010).

In: Global Climate Change … ISBN: 978-1-62081-288-4
Editors: A. C. DiStefano and M. Hurley © 2012 Nova Science Publishers, Inc.

Chapter 2

CLIMATE CHANGE ADAPTATION: FEDERAL EFFORTS TO PROVIDE INFORMATION COULD HELP GOVERNMENT DECISION MAKING*

United States Government Accountability Office

WHY GAO DID THIS STUDY

Climate change is a complex, crosscutting issue that poses risks to many existing environmental and economic systems, including agriculture, infrastructure, ecosystems, and human health. A 2009 assessment by the United States Global Change Research Program (USGCRP) found that climate-related changes—such as rising temperature and sea level—will combine with pollution, population growth, urbanization, and other social, economic, and environmental stresses to create larger impacts than from any of these factors alone.

According to the National Academies, USGCRP, and others, greenhouse gases already in the atmosphere will continue altering the climate system into the future, regardless of emissions control efforts. Therefore, adaptation—

* This is an edited, reformatted and augmented version of the Highlights of GAO-12-238T, a testimony before the Subcommittee on Oceans, Atmosphere, Fisheries, and Coast Guard, Committee on Commerce, Science, and Transportation, U.S. Senate, dated November 16, 2011.

defined as adjustments to natural or human systems in response to actual or expected climate change—is an important part of the response to climate change.

This testimony addresses (1) the data challenges that federal, state, and local officials face in their efforts to adapt to a changing climate, (2) the actions federal agencies could take to help address these challenges, and (3) federal climate change strategic planning efforts. The information in this testimony is based on prior work, largely on GAO's recent reports on climate change adaptation (GAO-10-113) and federal climate change funding (GAO-11-317). These reports are based on, among other things, analysis of studies, site visits to areas pursuing adaptation efforts, and responses to a web-based questionnaire sent to federal, state, and local officials.

WHAT GAO FOUND

As GAO reported in October 2009, challenges from insufficient site-specific data—such as local projections—make it hard for federal, state, and local officials to predict the impacts of climate change, and thus hard to justify the current costs of adaptation efforts for potentially less certain future benefits. Based on responses from a diverse array of federal, state, and local officials knowledgeable about adaptation, related challenges generally fit into two main categories: (1) translating climate data—such as projected temperature and precipitation changes—into information that officials need to make decisions and (2) the difficulty in justifying the current costs of adaptation with limited information about future benefits.

Federal actions to provide and interpret site-specific information would help address data challenges associated with adaptation efforts, based on responses to GAO's web-based questionnaire sent to federal, state, and local officials and other materials analyzed for its October 2009 report. In addition to several potential federal actions identified as useful by respondents to GAO's questionnaire, including the development of state and local climate change vulnerability assessments, GAO's 2009 report also contained information about the creation of a federal climate service. Specifically, about 61 percent (107 of 176) of respondents rated the "creation of a federal service to consolidate and deliver climate information to decision makers to inform adaptation efforts" as very or extremely useful. Respondents offered a range of potential strengths and weaknesses for such a service. For example, several respondents stated that a climate service would help consolidate information and provide a single information resource for local officials. However, some respondents to GAO's questionnaire voiced skepticism about whether it was feasible to consolidate climate information, and others stated that such a service would be too rigid and may get bogged down in lengthy review processes. GAO has not made recommendations regarding the creation of a climate service within the National Oceanic and Atmospheric Administration or any other agency or interagency body.

Federal strategic planning efforts could be improved for many aspects of the climate change enterprise. For example, GAO's October 2009 report on climate change adaptation concluded that, to be effective, related federal efforts must be coordinated and directed toward a common goal. This report recommended the development of a strategic plan to guide the nation's efforts to adapt to a changing climate, including the identification of mechanisms to increase the capacity of federal, state, and local agencies to incorporate information about current and potential climate change impacts into government decision making. Some actions have subsequently been taken to improve federal adaptation efforts, but GAO's May 2011 report on climate change funding found that federal officials do not have a shared understanding of strategic governmentwide priorities.

Chairman Begich, Ranking Member Snowe, and Members of the Subcommittee:

I am pleased to be here today to discuss federal efforts to provide climate data and services to decision makers. Climate change is a complex, crosscutting issue that poses risks to many existing environmental and economic systems, including agriculture, infrastructure, ecosystems, and human health. A 2009 assessment by the United States Global Change Research Program (USGCRP) found that climate-related changes—such as rising temperature and sea level—will combine with pollution; population growth; urbanization; and other social, economic, and environmental stresses to create larger impacts than from any of these factors alone.[1] According to the National Academies, USGCRP, and others, greenhouse gases already in the atmosphere will continue altering the climate system into the future, regardless of emissions control efforts. Therefore, adaptation—defined as adjustments to natural or human systems in response to actual or expected climate change—is an important part of the response to climate change.

Many federal entities manage climate change programs and activities. According to the Office of Management and Budget's June 2010 Federal Climate Change Expenditures Report to Congress, 9 of the 15 cabinet-level departments, along with 7 other federal agencies, received funding for climate change activities in fiscal year 2010.[2] In addition, entities within the Executive Office of the President, such as the Office of Scienceand Technology Policy, and federal interagency coordinating bodies, USGCRP, work together to ensure federal climate change activities are guided by the latest climate science. A September 2010 report by the National Academy of Public Administration, which was prepared for the National Oceanic and Atmospheric Administration (NOAA) and Congress, referred to this set of federal activities as the federal "climate change enterprise."[3]

Federal climate programs are shifting their focus to adaptation and climate services. Our October 2009 report on climate change adaptation found no coordinated national approach to adaptation, but our May 2011 report on climate change funding cited indications that federal agencies were beginning to respond to climate change more systematically.[4] About the same time as the issuance of our October 2009 report, Executive Order 13514 on Federal Leadership in Environmental, Energy, and Economic Performance called for federal agencies to participate actively in the Interagency Climate Change Adaptation Task Force.[5] The task force, which began meeting in Spring 2009, is co-chaired by the President's Council on Environmental Quality, NOAA, and the Office of Science and Technology Policy and includes representatives from more than 20 federal agencies and executive branch offices. The task force was formed to develop federal recommendations for adapting to climate change impacts both domestically and internationally and to recommend key components to include in a national strategy. In addition, USGCRP recently launched a national climate assessment designed to engage stakeholders in a process that builds on science, data, and information to help decision making. Individual agencies are also beginning to consider adaptation actions. For example, in May 2009, the Chief of Naval Operations created Task Force Climate Change to address the naval implications of a changing Arctic and global environment.

My testimony today addresses (1) the data challenges that federal, state, and local officials face in their efforts to adapt to a changing climate, (2) the actions federal agencies could take to help address these challenges, and (3) federal climate change strategic planning efforts. The information in this testimony is based on prior work, largely on our recent reports on climate change adaptation and federal climate change funding.[6] Our work was based on, among other things, analysis of studies; site visits to areas pursuing adaptation efforts; responses to a web-based questionnaire sent to federal, state, and local officials knowledgeable about adaptation; and interviews with such officials. A detailed description of our scope and methodology is available in each issued product. All of the work on which this statement is based was performed in accordance with generally accepted government auditing standards.

A Lack of Site-Specific Data, Such as Local Projections of Expected Changes, Can Challenge the Ability of Officials to Manage the Effects of Climate Change

As we reported in October 2009, insufficient site-specific data, such as local projections of expected changes, make it hard for federal, state, and local officials to predict the impacts of climate change, and thus hard for these officials to justify the current costs of adaptation efforts for potentially less certain future benefits.[7] Based on the responses by a diverse array of federal, state, and local officials knowledgeable about adaptation to a web-based questionnaire designed for that report, related challenges generally fit into two main categories: (1) translating climate data—such as projected temperature and precipitation changes—into information that officials need to make decisions and (2) difficulty in justifying the current costs of adaptation with limited information about future benefits.

The process of providing useful information to officials making decisions about adaptation can be summarized by the following:

- First, data from global-scale models must be "downscaled" to provide climate information at a geographic scale relevant to decision makers. About 74 percent (133 of 179) of the officials who responded to our questionnaire rated "availability of climate information at relevant scale (i.e., downscaled regional and local information)" as very or extremely challenging.
- Second, the downscaled climate information must be translated into impacts at the local level, such as increased stream flow. Some respondents and officials interviewed for our October 2009 report said that it is challenging to link predicted temperature and precipitation changes to specific impacts. For example, one federal official said that "we often lack fundamental information on how ecological systems/species respond to non-climate change related anthropogenic stresses, let alone how they will respond to climate change."

- Third, local impacts must be translated into costs and benefits, since this information is required for many decision making processes. Almost 70 percent (126 of 180) of the respondents to our questionnaire rated "understanding the costs and benefits of adaptation efforts" as very or extremely challenging.[8] As noted by one local government respondent, it is important to understand the costs and benefits of adaptation efforts so they can be evaluated relative to other priorities.
- Fourth, decision makers need baseline monitoring data to evaluate adaptation actions over time. Nearly 62 percent (113 of 181) of the respondents to our questionnaire rated the "lack of baseline monitoring data to enable evaluation of adaptation actions (i.e., inability to detect change)" as very or extremely challenging.

These challenges make it difficult for officials to justify the current costs of adaptation efforts for potentially less certain future benefits. A 2009 report by the National Research Council (NRC) discusses how officials are struggling to make decisions based on future climate scenarios instead of past climate conditions.[9] According to the report, requested by the Environmental Protection Agency and NOAA, usual practices and decision rules (e.g. for building bridges, implementing zoning rules, using private motor vehicles) assume a stationary climate—a continuation of past climate conditions, including similar patterns of variation and the same probabilities of extreme events. According to the NRC report, that assumption, which is fundamental to the ways people and organizations make their choices, is no longer valid.

Federal Actions to Provide and Interpret Site-Specific Information Would Help Officials Understand the Impacts of Climate Change and Available Adaptation Strategies

Federal actions to provide and interpret site-specific information would help address challenges associated with adaptation efforts, based on our analysis of responses to the web-based questionnaire and other materials analyzed for our October 2009 report.[10] The report discussed several potential federal actions that federal, state, and local officials identified as useful to inform adaptation decision making. These included state and local climate change impact and vulnerability assessments and the development of processes and tools to access, interpret, and apply climate information. In that report, we also obtained information regarding the creation of a climate service—a federal service to consolidate and deliver climate information to decision makers to inform adaptation efforts.

About 61 percent (107 of 176) of the federal, state, and local officials who responded to the web-based questionnaire developed for our October 2009 adaptation report rated the "creation of a federal service to consolidate and deliver climate information to decision makers to inform adaptation efforts" as very or extremely useful.[11] Respondents offered a range of potential strengths and weaknesses for such a service. Several said that a climate service would help consolidate information and provide a single-information resource for local officials, and others said that it would be an improvement over the current ad hoc system. A climate service would avoid duplication and establish an agreed set of climate information with uniform methodologies, benchmarks, and metrics for decision making, according to some officials. According to one federal official, consolidating scientific, modeling, and analytical expertise and capacity could increase efficiency. Similarly, some officials noted that with such consolidation of information, individual agencies, states, and local governments would not have to spend money obtaining climate data for their adaptation efforts. Others said that it would be advantageous to work from one source of information instead of different sources of varying quality. Some officials said that a climate service would demonstrate a federal commitment to adaptation and provide a credible voice and guidance to decision makers. In an announcement on February 8, 2010, the Department of Commerce proposed establishing a NOAA climate service. Though not yet established, information is available on the NOAA climate service website, including draft vision and strategic framework documents.[12] According to NOAA documents, such a climate service would provide a single, reliable, and authoritative source for climate data, information, and decision support services to help individuals, businesses, communities, and governments make smart choices in anticipation of a climate changed future.[13] A September 2010 report by the National Academy of Public Administration discusses the factors needed for a NOAA climate service to succeed—such as the designation of a lead federal agency to be the day-to-day integrator of the overall federal effort regarding climate science and services—and makes recommendations on how to achieve those factors.[14]

Other respondents to our questionnaire, however, were less enthusiastic about the creation of a climate service. Some voiced skepticism about whether it was feasible to consolidate climate information, and others said that such a system would be too rigid and may get bogged down in lengthy review processes. Furthermore, certain officials stated that building such capacity may not be the most effective place to focus federal efforts because the information needs of decision makers vary so much by jurisdiction. Several officials noted that climate change is an issue that requires a multidisciplinary response, and a single federal service may not be able to supply all of the necessary expertise. For example, one federal official stated that the information needs of Bureau of Reclamation water managers are quite different from the needs of Bureau of Land Management rangeland managers, which are different from the needs of all other resource management agencies and programs. The official stated that it seems highly unlikely that a single federal service could effectively identify and address the diverse needs of multiple agencies. Several respondents also said that having one preeminent source for climate change information and modeling could stifle contrary ideas and alternative viewpoints. Moreover, several officials who responded to our questionnaire were concerned that a climate service could divert attention and resources from current adaptation efforts by reinventing duplicative processes without making use of existing structures. The 2009 NRC report on informing decisions in a changing climate recommends that the federal government's adaptation efforts should be undertaken through a new integrated interagency initiative with both service and research elements but that such an initiative should not be centralized in a single agency.[15] Doing so, according to this report, would disrupt existing relationships between agencies and their constituencies and formalize a separation between the emerging science of climate response and fundamental research on climate and the associated biological, social, and economic phenomena. Furthermore, the report states that a climate service located in a single agency and modeled on the weather service would by itself be less than fully effective for meeting the national needs for climate-related decision support. The NRC report also notes that such a climate service would not be user-driven and so would likely fall short in providing needed information, identifying and meeting critical decision support research needs, and adapting adequately to changing information needs.

We have not made recommendations regarding the creation of a climate service within NOAA or any other agency or interagency body, although the provision of climate data and services will be an important consideration in future governmentwide strategic planning efforts, particularly in an era of declining budgets.

FEDERAL CLIMATE CHANGE STRATEGIC PLANNING EFFORTS COULD BE IMPROVED

Federal strategic planning efforts could be improved for many aspects of the climate change enterprise. Our October 2009 report on climate change adaptation concluded that, to be effective, related federal efforts must be coordinated and directed toward a common goal.[16] This report recommended the development of a strategic plan to guide the nation's efforts to adapt to a changing climate, including the identification of mechanisms to increase the capacity of federal, state, and local agencies to incorporate information about current and potential climate change impacts into government decision making. Some actions have subsequently been taken to improve federal adaptation efforts, but our May 2011 report on climate change funding found that federal officials do not have a shared understanding of strategic governmentwide priorities.[17] This report recommended, among other things, the clear establishment of federal strategic climate change priorities, including the roles and responsibilities of the key federal entities, taking into consideration the full range of activities within the federal climate change enterprise. In other reports, we also noted the need for improved coordination of climate-related activities. For example, our April 2010 report on environmental satellites concluded that gaps in satellite coverage, which could occur as soon as 2015, are expected to affect the continuity of important climate and space weather measurements.[18] In that report, we stated that, despite repeated calls for interagency strategies for the long-term provision of environmental data from satellites (both for climate and space weather purposes), our nation still lacks such plans.

Of particular importance in adaptation are planning decisions involving physical infrastructure projects, which require large capital investments and which, by virtue of their anticipated lifespan, will have to be resilient to changes in climate for many decades. The long lead time and long life of large infrastructure investments require such decisions to be made well before climate change effects are discernable. Our ongoing work for the Senate Committee on Environment and Public Works Subcommittee on Oversight and Subcommittee on Transportation and Infrastructure will explore this issue by reviewing the extent to which federal, state, and local authorities consider the potential effects of climate change when making infrastructure investment decisions.

Chairman Begich, Ranking Member Snowe, and Members of the Subcommittee, this concludes my prepared statement. I would be happy to respond to any questions that you or other Members of the Subcommittee may have.

End Notes

[1] USGCRP coordinates and integrates federal research on changes in the global environment—including climate change—and their implications for society.

[2] Office of Management and Budget, Federal Climate Change Expenditures Report to Congress (June 2010). See http://www.whitehouse.gov/sites/default/files/omb/assets/legislative _reports/ FY2011_Climate_Change.pdf

[3] Panel of the National Academy of Public Administration, *Building Strong for Tomorrow: NOAA Climate Service*, a report prepared for Congress, the Department of Commerce, and NOAA (Sept. 13, 2010).

[4] GAO. *Climate Change Adaptation: Strategic Federal Planning Could Help Government Officials Make More Informed Decisions*, GAO-10-113 , (Washington, D.C.: Oct. 7, 2009), and *Climate Change: Improvements Needed to Clarify National Priorities and Better Align Them with Federal Funding Decisions*, GAO-11-317 , (Washington, D.C.: May 20, 2011).

[5] For more information about the Interagency Climate Change Adaptation Task Force, see http://www.whitehouse.gov/administration/eop/ceq/initiatives/adaptation.

[6] GAO-10-113 and GAO-11-317 .

[7] GAO-10-113 .

[8] The number of respondents varies because some officials did not respond to certain questions.

[9] National Research Council of the National Academies, Panel on Strategies and Methods for Climate-Related Decision Support, Committee on the Human Dimensions of Global Change, *Informing Decisions in a Changing Climate* (Washington, D.C., 2009).

[10] GAO-10-113.

[11] GAO-10-113 .

[12] For more information about the NOAA Climate Service, see http://www.noaa.gov/climate. html. A range of climate information is presented at www.climate.gov, NOAA's Climate Services Portal.

[13] The Department of Defense and Full Year Continuing Appropriations Act, 2011 prohibited any funds appropriated in the act to be used to implement, establish, or create a NOAA Climate Service as NOAA had previously described it during fiscal year 2011.

[14] Panel of the National Academy of Public Administration, *Building Strong for Tomorrow: NOAA Climate Service*, a report prepared for Congress, the Department of Commerce, and NOAA (Sept. 13, 2010).

[15] USGCRP's September 30, 2011 Draft Strategic Plan reflects elements of these NRC recommendations.

[16] GAO-10-113 .

[17] GAO-11-317.

[18] GAO. *Environmental Satellites: Strategy Needed to Sustain Critical Climate and Space Weather Measurements*, GAO-10-456 , (Washington, D.C.: Apr. 27, 2010). For another example of the need for improved strategic planning, see *Climate Change: A Coordinated Strategy Could Focus Federal Geoengineering Research and Inform Governance Efforts*, GAO-10-903 , (Washington, D.C.: Sept. 23, 2010).

INDEX

#

A

B

C

D

E

F

J

K

L

M

N

O

P

Q

R

S

T

U

V

W

Y

Z